What Is Science?

An Interdisciplinary View

Klaus Jaffe

T0127879

UNIVERSITY PRESS OF AMERICA,® INC.
Lanham • Boulder • New York • Toronto • Plymouth, UK

Copyright © 2010 by
University Press of America,® Inc.
4501 Forbes Boulevard
Suite 200
Lanham, Maryland 20706
UPA Acquisitions Department (301) 459-3366

Estover Road
Plymouth PL6 7PY
United Kingdom

Library of Congress Control Number: 2009930542
ISBN: 978-0-7618-4673-4 (paperback : alk. paper)
eISBN: 978-0-7618-4674-1

Contents

Warning

1- The intent of this book is not to be a scholarly, complete, and exhaustive treaty on science and its methods. Rather, I want to analyze both old and novel ideas concerning the subject with a broad transdisciplinary perspective and to explore new vistas, aiming to develop a unified interdisciplinary definition of science. Many topics discussed in this work are only superficially analyzed. They can be explored in depth elsewhere and the reader is advised to do so. Here, I provide a broad holistic view, rather than engage in a detailed reductionist analysis of science. This approach has important limitations, as many relevant details have to be omitted, but it allows for levels of generalization and abstraction that are not achieved by explorations of a single discipline. The cost of an extensive and wide analysis is a limited depth of knowledge. As stated in popular wisdom, "We can know very much about very little or very little about very much". Here, I chose the second option, but acknowledge the relevance of the first.

2- The term science, as used here, is based on the legacy of many scientist, among which Galileo Galilei seems to be the most prominent, in which science is defined as the epistemology that subordinates theory to empirical data and experimental results. I am aware that many authors equate science with epistemology. Here, I follow the conviction that not all epistemological systems are scientific and they should not be confused with the modern experimental version of science. Thus, the term "science" refers to the "Galileanic" version, while the term "sciences" refer to other epistemologies (branches of philosophy that deal with the nature, origin, and scope of knowledge).

3- The citations of authors and previous relevant works are not in accordance with traditional scientific rules. Text in this work have been designed to help the reader search the Internet by introducing a relevant phrase, name or word into an intelligent search engine (such as Google or Coperic) to initiate a new exploration for knowledge.

Preface: In Search of a Multidimensional Heuristics

What is nature? What is reality? What is life? What is consciousness? How do our minds work? How do we acquire new knowledge? How do we best explore reality? How do we increase our knowledge? Where does technology come from? What is science? These are questions normally posed by philosophers, theologists, and broad thinking intellectuals. The term heuristic (technique to discover new knowledge) covers all of these endeavors. Scientists rarely bother to answer these questions. They argue that they are busy answering real and concrete questions, an activity which does not leave time for abstract queries. Yet, the answers to these questions are important in order to obtain a glimpse into the future of humanity. The type of person that answers these questions is also important. Scientists have direct experience practicing science and acquiring new knowledge through the exploration of reality that philosophers normally lack. Thus, scientists and philosophers have divergent views and these differences matter, as they will affect the answers to the posed questions. The mainstream philosophy of science and our understanding of the workings of science have been overwhelmingly exercised by philosophers and not by scientists. The time has arrived to alter this tendency.

Physicists are among the few scientists who have attempted, with certain success, to take the study of science out of the hands of philosophers and humanists. Biologists and scientists engaged in interdisciplinary research have rarely attempted to meddle in the discussions between philosophers and physicists regarding the definition of science and its workings. Here, I will attempt to disrupt this monopoly on the analysis of the nature and practice of science. I hope that this attempt will shed new light on the mysterious and dark areas of our knowledge landscape. Complexity theory, evolutionary biology, and comparative ethology have allowed for the development of analytical tools and concepts that may facilitate a better understanding of the adaptive value of science by clarifying ancient contradictions related to science and its methods. This might allow us to transcend the work of many brilliant minds, such as a number of classic Greek philosophers, Immanuel Kant (*Critic of Pure Reason*), Ludwig Wittgenstein (*Tractatus Logico-Philosophicus* and *Philosophical Investigations*), and many others.

People that have been in many diverse environments that provide them with a broad and deep view of the world and that have the genius to produce fundamentally new ideas have rarely developed the linguistic skills required to transmit their ideas to a broader public audience. Scientific geniuses with extraordinary communication skills, such as Galileo, Darwin, Huxley, and Tyndall, are rare. Yet, when the new ideas are not based on linear logic subjacent to language, the rare coincidence of scientific genius and communication skills is even less likely. Individuals with a genetic makeup and experience that enables the acquisition and development of superb language skills which favor the construction of long linear chains of ideas will likely distance themselves from intellectual arguments that favor explorations of an intangible, non-linear web-like, complex, multidimensional, non-adaptive, diffuse reality. Yet complexity and multidimensionality characterize the present frontier of modern science.

May this preface serve as a warning to the reader who expects to penetrate amazing new worlds without any intellectual effort on his/her side. At the same time, may this comment serve as an anticipatory excuse, as I would love to have written this book using poetry. Instead, I compensate my linguistic limitations using images that I hope will help convey the message. The use of images, however, has its shortcomings. Images might be very powerful messengers of ideas, but they do not always convey the exact intended meaning and they may be interpreted in different ways by different people. The images will serve here as a metaphor to help convey a message, but intellectual effort is required from the reader, so that new ideas can be brought to life.

The present work aims to open new avenues of discussion and suggest new research routes, rather than to convey the impression of a finished work. The emergence of science was certainly the most important event for humanity in the last millennium and the end of the scientific revolution is still not in sight. I hope to impart a better understanding of this historic process in order to increase the odds of its potential benefits to humanity.

This endeavor, however, will require an understanding of the structure of reality, its interplay with our consciousness and the dynamics of acquiring new knowledge. The space in which these interactions occur will be called multidimensional, for lack of a better name. Our aim here will be to explore a heuristic of the multidimensional reality of which we are also a part.

Acknowledgements

Thanks are due to Benjamin Scharifker, Juan Torres, Joaquin Medin, Virgilio Lew, Carlo Caputo, James Randi, Guillermo Mascitti, Joaquin Lira, Paola Bressan, Robert Axtell, Rodolfo Jaffe, Violeta Rojo, Werner Jaffe, Manuel Bemporad, Tyra Bacon, Eva Melinda, and an innumerable list of anonymous referees for insightful discussions, productive ideas, and/or suggestions and corrections that shaped this book. I can not avoid mentioning the following landscapes that inspired parts of this book: Guadeloupe in the French Antilles; Marahuaca-tepuy and Hato La Fe in Venezuela; Big Island in Hawaii, Tarija and the surrounding Altoplano in Bolivia; Salta in Argentina; Rio Miranda in the Pantanal, Brazil; and of course Caracas, Venezuela

Introduction

The history of science and the philosophy of science are mostly based on analyses of the histories of physical science. Physical science indicated the modest place in the cosmos that is occupied by our planet and revealed the striking laws of nature that led to the construction of the most potent means of destruction that humans have ever built: the atomic bomb. However, other weapons of mass destruction are now competing with the efficiency of the atomic bomb in their deadly endeavor, revealing advancement of other scientific disciplines that have been less apparent to the public. Thus, physics is not the sole scientific field and the importance of the physical sciences could diminish in the future compared to more modern disciplines, such as biology, economy, psychology, and sociology.

The philosophy of science, through the analysis of scientific phenomena viewed through the lens of physical science, has produced important insights into the working of science. It has suggested the existence of the scientific method (Mario Bunge: *Epistemologia*); that scientific progress is punctuated rather than continuous (Thomas Kuhn: *The Structure of Scientific Revolutions*); that disproving a scientific theory is much more important for scientific progress than supporting it (Karl Popper: *Logik der Forschung*); that scientific dynamics can not be planned ahead or controlled from the outside (Paul Feyerabend: *Against Method*); and that these dynamics are driven by *The Methodology of Scientific Research Programmes* (Imre Lakatos: *Proofs and Refutations*). These deep insights are certainly relevant to all disciplines of science.

The grounding of our vision of science on physics might have been the simplest and most efficient way to study the social dynamics of scientific phenomenon, but it is time to go further and re-analyze the workings of science and its effect on society in a broader and more interdisciplinary context. More specifically, this analysis should include modern insights from evolutionary biology. This is the aim of the present work.

Many of the newer scientific disciplines have not been recognized as being completely scientific. Many of them base their source of knowledge on broad verbal descriptions of poorly understood phenomena. A better understanding of the dynamics and the workings of science might facilitate the maturation of those disciplines. This is one of the explicit aims of this book. Further development in sociology, economy, and psychology with firmer roots in science will benefit those disciplines in profound ways as it will allow us to understand humans and their societies more completely. The benefits of such results are enormous and unimaginable. Thus, a walk on the path to a better understanding of newly developing scientific disciplines seems to be well justified, even if we stumble over theoretical, moral, conceptual, or ideological obstacles. It is our hope that any shortcoming of the present book will serve as a motivational force for others to improve our understanding of the workings of science and to obtain a better view of the social dynamics of science.

In tackling this subject, we should always keep in mind that we lack a perfect approach to study science. A general exploration necessarily limits in-depth analysis; in-depth analysis necessarily hinders a general view of the subject. The multidimensional nature of reality allows it to be analyzed by many different approaches, but the limited space of any book only permits presentation of a tiny fraction of it. Thus, I ask the reader to judge an analytical approach to reality by its heuristic merits and its capacity to achieve an interdisciplinary synthesis, rather than on its completeness and its rigorous presentation of details. Of course, my personal experience with science will permeate the visions explored here. I beg the reader to tolerate my shortcomings but offer a rough ride through important philosophical and practical questions.

Chapter One

The Ascent of
Homo sapiens scientiarum

"The smallest of events presupposes the inconceivable universe, and reversibly, the universe needs the smallest of events"

—J.L. Borges

NEGENTROPIC PROCESS OF EVOLUTION:
A WORLD FOR THE LUCKY

It is well accepted that all living organisms on our planet, including *Homo sapiens*, are the product of biological evolution. Our present insights into the dynamics of biological evolution indicate that it is propelled by three basic forces or systems:

1. Transmission of information (from parent to offspring through genetic inheritance)
2. Production of variation (through mutation and genetic recombination)
3. Natural selection (with a healthy dose of chance)

Biological evolution, working on these three principles, is certainly a complex process. It involves the action and interaction of many different components, and in addition, its dynamics are irreversible. From a purely mathematical standpoint, an analytical search for a putative optimum combination of genetic traits by evolution of a complex assemblage of genes, mimicking real organisms, is not possible at present, and may even be impossible in principle. In addition, we know that stochastic (or random) and chaotic processes influence the final outcome of natural selection. Thus, a

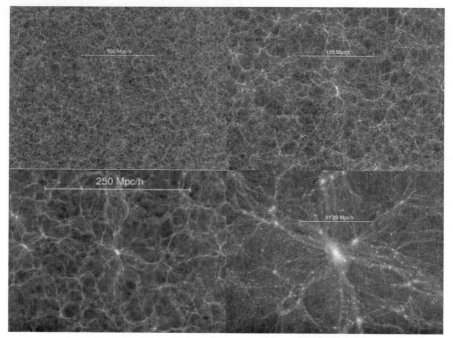

Figure 1.1.   Patterns of the Cosmos at four different scales from simulation of the formation, evolution and aggregation of galaxies and quasars. Published in the journal Nature of June 2, 2005.

complete understanding of biological evolution is not trivial. Natural selection does not always select the optimal available solution, but often favors the *luckiest* one, i.e., evolutionary advancements are often determined by chance in addition to the constraints of history, geography, and the dynamics of a particular biological and physical process, rather than by the parsimony and efficiency of the selected alternative. Chance and adaptation both determine whether a species establishes itself among the living. That is, we live in a world for the lucky.

The practical outcome of the survival of the fittest is operationally equivalent to a drive for *transcending*. In other words, the main *raison d'etre* of all living beings is to achieve permanence on the planet, as long as possible, be it through its own survival or that of its descendants or relatives. Expressed in classical evolutionary terms, the adaptive value of any behavior or feature of an organism is its increase in fitness, which can be achieved by increased odds of survival, reproduction, or both. This is achieved through three simple dynamic features or algorithms:

1. Transmission of information (mainly vertical transmission by genetic inheritance, although horizontal transmission of genetic and cultural characters also occurs)
2. Variance (mainly mutations and genetic recombination caused by sexual reproduction)
3. Natural selection (affecting genes, gametes, organisms, groups, populations, and species; the most important level of selection is at the level of organisms that interact in populations with their environment).

The mechanisms of biological evolution, including a series of algorithms, behaviors, structures, and devices, have themselves evolved to better transcend present life. They have accumulated in most extant (alive at present) organisms, which has allowed for interesting interactions. As the number and sophistication of such devices increases, the scope for transcending widens. More sophisticated tools are required to effectively control an increasingly unpredictable environment. We can reasonably assume that biological evolution will eventually produce ever more sophisticated organs that provide better means to increase their success in the battle for survival. In other words, complex sophisticated devices are not likely to evolve into simple undifferentiated ones (although this might occur, among parasites for example), but are very likely to have evolved from simpler ones. Thus, biological evolution (and cosmological evolution) shows a clear direction in time, starting from the simple and undifferentiated, and leading to the complex and sophisticated. This tendency is called thermodynamic negentropy. The opposite tendency, leading from the complex and organized to the simple and disorganized, is called entropy.

These negentropic tendencies are not always evident in evolutionary processes; but, they can be clearly observed if we have the right perspective in time.

NATURAL HISTORY OF SCIENCE

It is easy to loose perspective when analyzing time windows that are much greater than our life span. Our brains have not adapted to think in these time frames. Thus, it would be helpful to have a bird's eye view of history, by expanding our analysis as deep as possible in time, in order to analyze the past millennium. I will try to do so by using different time windows to analyze our last millennium (see Figure 1.2).

In the largest imaginable time window, comprising tens of millions of millennia, our millennium is not visible. This window serves to place formation

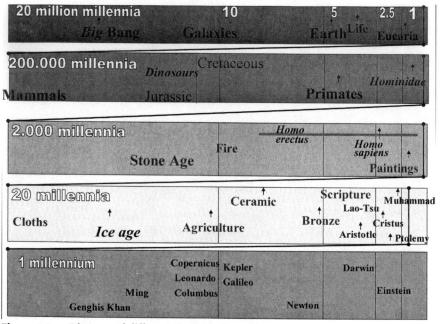

Figure 1.2.   Diagram of different time windows.

of the planet Earth and the emergence of life in perspective. If the event that created our present universe, the Big Bang, occurred some 14–15 million millennia ago, then the development of life on the planet earth represents about 25% of the history of our universe.

One percent of this large time window produces a second time window spanning 200,000 millennia. This period includes the evolution of extant animals and plants and the emergence of primates and hominids. Analyses of this time scale indicate a discontinuous evolutionary process: Geological catastrophes, such as those that created the Jurassic/Cretaceous boundary, mark the onset of the evolution of new life forms.

One percent of this last time window produces a third time window, spanning 2,000 millennia, in which human innovation becomes apparent. Stone tools were used by our predecessors and fire was domesticated before the formation of the oldest known fossils of *Homo erectus* and the emergence of *Homo sapiens*. This suggests that creativity, technological development, and inventiveness are features that were possessed by at least some of the australopithecine and other hominids long before *Homo sapiens* appeared. Human creativity developed slowly over the last 5,000 millennia. Interestingly, with these time scales, a number of extinction's of other human-like creatures (*Homo erectus,* for example) are observed. The extinction of other hominids and mammals (the mastodons, for example) were probably caused by the increased technical powers of *Homo sapiens,* such as the use of traps, spears, and projectiles.

A fourth time window, spanning one percent of the former, focuses on our past using a 20 millennia lens. This lens shows that human creativity (and the extinction of fellow creatures on our planet) continues to accelerate exponentially. Cloths used to protect the human body from the environment were woven before the last ice age. The end of this ice age promoted the domestication of plants and animals, leading to the development of agriculture over 10 millennia ago. The development of ceramics, the wheel, iron, bronze, steel, and scriptures followed at ever shorter intervals. A fundamental human advance during this period was the development of writing. The use of writing permits a more direct view of the thought processes of our ancestors. Homer, Lao-Tse, Buddha, Aristotle, Christ, Ptolemy, Muhammad, and many other thinkers are accessible from their scripts or transcripts of their talks.

Five percent of this 20 millennia time window contains our last millennium. The main feature of the last millennium seems to be the emergence of the scientific method. Science, defined as the method which subordinates theory to experimental results, was pioneered by Galileo among others, and has been widely accepted as a superior form of thought during the last three centuries. Another interesting aspect of this last millennium is a shift of creative power from China to the West. The nationalistic and xenophobic policies of the Ming dynasty (started by Chu-Yuan-Chang in 1368) hindered progress, such as that of Admiral Zheng He. The efforts of Admiral He in maritime exploration and shipbuilding between 1405 and 1433 proved so successful in advancing knowledge, technology, and power, including the discovery of the rest Asia and Africa, and possibly America, for the Chinese, that the Chinese Emperor ordered the destruction of Admiral He's fleet. The abandonment of maritime explorations disrupted China's international commerce and arrested the recent exponential expansion of its creativity. This burst of creativity was motored by exposure to foreign ideas and technology, catalyzed by the expansion of maritime transport.

A similar historic pause of a few centuries in the growth of creativity occurred also in Christian Europe, spanning from the first millennium AC (approximately after the deaths of Ptolemy and the Roman Emperor Julianus) to the Renaissance, overlapping with a burst of creativity in the Islamic cultures of that time.

There has been an exponential expansion of human mental capabilities and the last millennium shows no signs of the expansion reaching a plateau. Analysis of the evolution of other animal species, however, suggests that all exponential evolutionary developments of behavioral and morphological traits eventually stop. Even human creativity may start to do so in the third millennium AC.

The most important consequence of the appearance of the experimental scientific method, which allowed for the profusion of the natural science, was the accelerated development of technology. This triggered the "Industrial Revolution" and caused exponential economic growth. History teaches us that science defines victory. The battles of Austerlitz, Trafalgar, and Hiroshima were won by those who managed the most advanced scientific knowledge with the greatest efficiency.

This wide historic sweep shows that the emergence of science, as well as the emergence of any biological feature, is the product of evolution and must be viewed in a logarithmic timescale in order to grasp its historical context. In a similar way, the consequences of the emergence of science unleashed a series of events that can only be understood with a wide view along a logarithmic timescale.

CREATIVITY AND IMAGINATION

The protagonists of the natural history of science are certainly the forces that create novelty. What are these forces and how do they work? We do not know much about the detailed mechanisms that promote creativity, but we do know that technological inventiveness accelerated conspicuously after the emergence of science.

Focusing on the last four thousand years of our history, we notice at least three different spurts of imaginative creativity with different subjacent characteristics (see Figure 1.3). A period of creative thinking centered on ancient Greece, founded on the knowledge accumulated by Egyptian and Persian scholars dominated the period of the second millennium BC. This period was followed by one in which creativity was focused on developing new religions and ways to promote mystical thinking. It is only with regard to the last 400–500 years that we can speak properly about science.

This last period of creativity, which was centered on science, was vastly more productive in terms of ideas, technology, and population growth, than any period before it. Of course, it is difficult to know if the scientific method was independently discovered beforehand during periods where it could not sprout. If we summarize the main advancements of human knowledge and technological abilities over the last 4 million years and compare them with creative innovations over the last 400 years (see Figures 1.4 and 1.5), the particularities of the effects of science may be evidenced. Although in both time windows, the working of creative forces is evident, it is only after the ap-

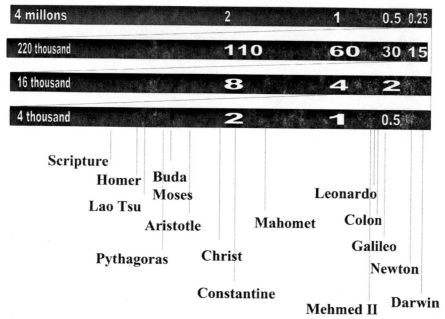

Figure 1.3.   Diagram of a historic time windows.

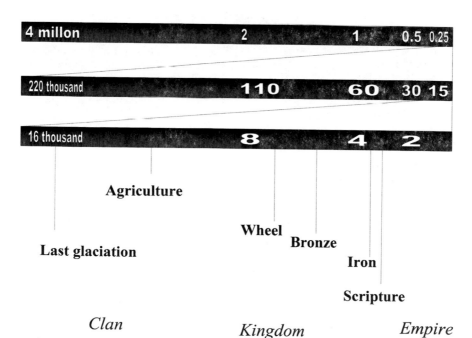

Figure 1.4.   Diagram of a historic time windows.

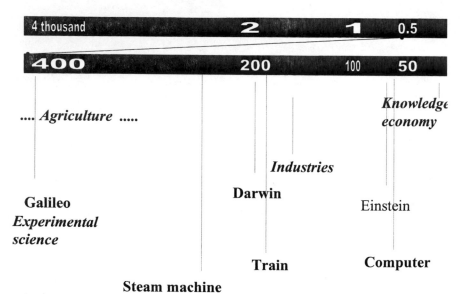

Figure 1.5.   Diagram of a historic time windows.

pearance of science that technological progress busts into exponential growth.

These insights facilitate the analysis of questions about the nature and uniqueness of science. Evidently, humans were creative and imaginative in Sumerian times. In ancient Greece, they produced astonishing intellectual feats and developed important and long lasting technologies. But, the frequency of these new inventions was sparse and the speed by which humanity invented and created new devices was much slower compared to modern times following the Industrial Revolution.

What are the differences between the wise ancient Greek geniuses and modern scientific scholars? Is this difference relevant to the differences in technological progress? Before answering these questions, we must resolve and clarify some other issues and concepts.

ORIGINS OF KNOWLEDGE: THE DRIVE TO KNOW

All organisms have developed algorithms or behavioral repertoires that determine their interactions with the environment in ways that increase their odds of survival and augment their reproductive success. Such algorithms have been selected through biological evolution and are encoded in the genomes of organisms. Accumulation of such knowledge by biological evolution has occurred, for example, through the selection of genes with successful mutations. Another important route for the accumulation of knowledge is through incorporation of foreign DNA into the host genome. A famous case of this 'sharing of DNA' is the symbiosis of unicellular organisms that led to the formation of modern cells or eukaryotes (complex cells with a differentiated nucleus). Eukaryotes were formed by the fusion of unicellular organisms, each with its own genome that was rich in information. Some of the organelles inside a eukaryotic cell are still recognizable as primitive bacteria. These organelles provide functions that synergize with other organelles, joining the accumulated knowledge acquired through two independent evolutionary processes.

Another way that unicellular organisms combined accumulated information was by aggregating to form multi-cellular organisms, in such a way that their original architecture and genomes remained intact. The emergence of multi-cellular organisms allowed individual cells to specialize in certain functions, including gathering, storing, and analyzing new information by aggregating into special tissues or organs in the organism.

More sophisticated organisms have more elaborate organs, behaviors, and systems that help them accumulate knowledge. Proteins receive signals in the bacteria's membrane and the cell plasma accumulates metabolites in response. Some species of fish living in muddy waters receive and process eclectic stimuli and remember the kind of signals that correlate with palatable prey. A tick uses its genetic memory which produces behavioral algorithms that instruct it to grab to its victim when carbon dioxide vapors and lactic acid reach its receptors. All living beings that interact actively with their environment, such as animals, require a device that allows them to filter the relevant stimuli from the environment in order to simplify the incoming information, to build a simplified version of the surrounding world, to place themselves in that world, and to explore the possible outcomes of different actions in that "virtual" world before taking risks in real-life. Some animals have achieved these tasks through the development of a central nervous system. Thus, before humans existed, evolutionary forces drove development of a powerful tool: our mind, that allows us to build models and explore our surroundings. This tool has been perfected in some social species by the emergence of culture, a socially based set of rules which allow for synergy between various interacting minds with complementary world experiences.

Mechanisms for accumulating knowledge and systems to track successful solutions do not seem to be sufficient for an organism to accumulate knowledge successfully. A drive to know is also required. This drive, in most higher organisms, seems to be especially strong when the animal is young. To that end, mutations present in some higher primates, especially humans, which maintain certain juvenile characteristics until old age seem to explain a hunger for knowledge among some human adults. The tendency to maintain juvenile characteristics in adult stages of life is called neotenism.

EMERGENCE AND EVOLUTION OF THE HUMAN MIND

As stated by Flinn, Geary, and Ward in their article *Ecological dominance, social competition, and coalitionary arms races: Why humans evolved extraordinary intelligence* (see Evolution and Human Behavior Volume 26, Pages 10–46): "*Human cognitive abilities are extraordinary. Our large brains are significantly modified from those of our closest relatives, suggesting a history of intense natural selection. The conditions favoring the evolution of human cognitive adaptations, however, remain an enigma. Hypotheses based on traditional ecological demands, such as hunting or climatic variability, have not provided satisfying explanations. Recent models based on social problem solving, linked with ecological conditions, offer more convincing scenarios. But it has proven difficult to identify a set of selective pressures that would have been sufficiently unique to the hominid lineage. What was so special about the evolutionary environments of our ancestors that caused them, and them alone, to diverge in such astonishing ways from their close relatives and all other life forms? Richard Alexander proposed a comprehensive integrated explanation. He argued that as our hominid ancestors became increasing able to master the traditional "hostile forces of nature," selective pressures resulting from competition among conspecifics became increasingly important, particularly in regard to social competencies. Given the precondition of competition among kin- and reciprocity-based coalitions (shared with chimpanzees), an autocatalytic social arms race was initiated, which eventually resulted in the unusual collection of traits characteristic of the human species, such as concealed ovulation, extensive biparental care, complex sociality, and an extraordinary collection of cognitive abilities. We term this scenario the "ecological dominance–social competition" (EDSC) model and assess the feasibility of this model in light of recent developments in paleoanthropology, cognitive psychology, and neurobiology. We conclude that although strong or direct tests are difficult with current data, Alexander's model provides a far-reaching and integrative explanation for the evolution of human cognitive abilities.*"

The proposition that the human mind was shaped by biological evolution, and that this evolution is responsible for human aggregate phenomena, providing the *Volksgeist* of an era, is not new. Georg Wilhelm Friedrich Hegel,

Auguste Compte, and Karl Marx dwelled on this possibility and made it central to their thoughts. We now know that several evolutionary stages and transitions have occurred during the evolution of Hominids. Even during the approximately 50,000 years spanned by the Neolithic man, biological evolution has continued to shape our mind. Sequencing of the genomes of humans and other animals indicate that numerous features, many related to the brain, evolved very recently or may still be in the process of evolving.

There have been important milestones in the evolution of our minds. One was the emergence of the subjective: The recognition that the self is part of the whole, but different from it as well. The introduction of the ego in the models elaborated by our minds clearly improved their accuracy and efficiency. This is thought to represent the cornerstone of human evolution (the discovery of "self" might have occurred in other non-human animal species also). Other milestones include the acceptance of logical rules; construction and use of mental models; inclusion of subjects in the model; the ability to have empathy and to imagine oneself inside the skin of others; the awareness that others know that I know; analysis of actions of a subject and other individuals through models; formalization of these phenomena and the taxonomic description of the world; analytical formalization of models using mathematics; and the use of computers to produce numerical models and perform simulations in virtual worlds.

These insights imply that human cognitive capabilities have emerged and evolved as products of human evolution as necessary adaptive features. The exact path of this development will remain unclear for many years to come, but a sketch that could provide some heuristically useful ideas is presented in Figure 1.6 which depicts the emergence of science in a schematic way.

Language, social contacts, and communication in general have been especially important in the evolution of the human brain and for the development of our mental capacities. These features consolidated in our mind long ago, even before the emergence of humans, and shaped the way rational thoughts are build. However, I will not discuss this evolution in detail. Regardless of the precise route by which evolution molded our minds, the ability to model and visualize an abstract and simplified version of reality in our minds has been fundamental to our development.

One consequence of the way the human mind emerged is that it has limitations, as does any functioning unit. It has slowly evolved to reflect the needs and constraints of the macro-physical world with which humans interact. For example, our nervous system has adapted to cope with a small range of gravitational force that is found on our planet and has not adapted to the gravitation that is encountered by airplanes, on space stations, or even on a rough

Human adaptation   Cognitive feature

       Human ability   External contact

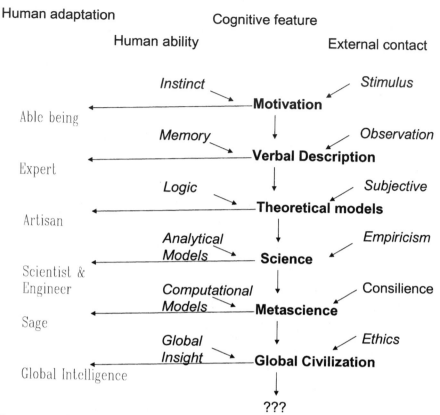

Figure 1.6.   Sketch of the evolution of human cognitive capabilities.

sea. We are familiar with a small window of frequencies of electromagnetic radiation (visible light) and a small range of object size (those visible with our eyes). In a sense, our mind mirrors the world that an organism had to interact with in the past (Konrad Lorenz: *Die Ruckseite des Spiegels*) as it was the best way to model the surrounding environment successfully.

Our mind has profited and still profits from evolutionary interactions with other minds, especially those of other humans. But, the minds of other species have also influenced us. Progress in human evolution can be viewed as the development of new symbioses between domesticated plants and animal and ourselves. The establishment of agriculture and cattle breading produced changes in human behavior and affected human culture in fundamental ways. The adoption of agriculture and the settlement of human populations was a fundamental step of our civilization. This symbiosis with plants and animals also affected the characteristics of domesticated organisms.

EVOLUTION OF RATIONALITY: THE MODELING MIND

To grasp objective reality is not a simple task. As Hermann Weyl (1885–1955) wrote in his *Philosophy of Mathematics and Natural Science,* "The objective world simply is, it does not happen. Only to the gaze of my consciousness, crawling along the lifeline of my body, does a section of this world come to life as a fleeting image in space which continuously changes in time."

Rational thought has evolved as an adaptation to grasp the objective reality in which we live and with which we interact. Rational thought is not restricted to humans. The design of a hunting strategy by a lion or an escape route through canopies by a monkey requires perception, mathematic reasoning, and planning skills that might be lacking in many mentally handicapped humans.

In the Stone Age and probably long before, human minds recognized certain regularities in their environments. The mind was able to produce abstractions of reality. An old evidence of this ability, which lies in the recognition and use of basic physical and biological laws, can be found in ancient cave paintings, some of which may have been painted over 70,000 years ago. These paintings (Figure 1.7) show exquisite abstractions of animals which perfectly convey information, including the animal species and its behavior, with a few lines.

Cave drawings suggest that an important step in the evolution of human rationality was to imagine an abstraction or "model" of the world, at least in terms of the immediate environment, which included game animals. This modeling allowed humans to teach, communicate, draw, and plan, which allowed for hunting in groups. The ability to perform finely tuned cooperative hunting, together with the rational design of hunting tools, might have given humans the upper hand in hunting beasts in nature.

From then on, mental models grew and increased in complexity, by creating and mastering language, grammar, religion, and world views.

A further step achieved by the oldest philosophy known to us, which is from Sumer in ancient Mesopotamia, exaggerated the importance of mental models. Sumerians assigned more importance to models than to reality. This is reflected in the fact that the verbalized models of reality become more important than reality itself. The words assume power that might change reality, at least as perceived by the verbalized models of reality in the minds of humans. The word creates reality, or, as stated in the Bible, at the beginning was the word. Sumerians, as well as many present day humans, assumed that the word of God was sufficient to create lives and beings in general. The verbally formulated law decreed by human Gods and kings constructed reality.

Figure 1.7. Paintings from the Cave "Barranco de la Valltorta" in Provincia Castellón, Spain.

Yet, even in ancient Sumer, some skeptical humans resisted believing only in words and drew wisdom from experience, observation, and experiments. It was not until the time of Galileo that experimentation was consciously placed above verbal models and human thoughts. This marked the real beginning of science as a cognitive driving force.

A model, conceptualized as a thinking tool, is a device that allows us to test the consistency of a set of assumptions and to explore the consequences of their possible interactions. In the visual example of modeling given below, the sphere on the right can be built with the components provided on the left. Yet, in order to test and explore the properties of this sphere, such as its dynamics when rolling or its floating properties, it is not enough to know and analyze the properties of the component parts. Many properties of the sphere will "emerge" from the interactions of the component parts. Thus, "modeling" in the sense of constructing an object in virtual space, allows us to explore some of these emerging properties with much less effort than would be required to build the sphere with real components (Figure 1.8). In this sense, models are thinking tools that allow thoughts to be created and to be put to work. Thus, these models guide our future actions and reduce future risks and costs.

The world is complex. Each of us develops, with varying degrees of success, mental models to help us come to terms with that complexity. These mental models, following Kenneth Craik (*The Nature of Explanation*), are a way of thinking or an internal manipulation of representations of how the

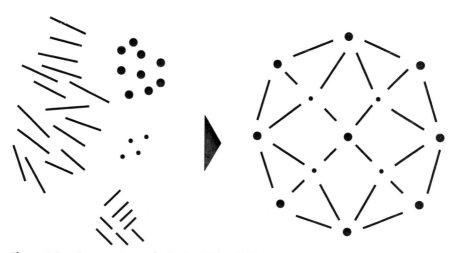

Figure 1.8.   **Emergent complexity by Alida Ribbi.**

world works. As we are not intelligent enough to cover all the details of actual reality in our imaginations, we have to work with simplified representations or models. Thus, good models provide improved insight into the behavior of vastly more complex real systems.

THE LIMITS OF PERCEPTION AND THOUGHT

One important insight of modern thought and especially of modern science is that our perceptions are very limited, distorted, and often lead us to errors. We are blind to high frequency light (above ultraviolet) and low frequency radiations (infrared and radio); we are deaf to ultrasound and very low frequency sound; we are anosmic to a large variety of volatiles and odors; we are incapable of comprehending infrequent events of great impact and compare them to the risks associated with very probable events of low impact; we insist on assigning causal relationships to random events; and we ineptly analyze a world in which we feel we are the main player. Simply put, we are not very objective in our rational views. These limitations mold our minds and restrict our thoughts, especially when analyzing aspects of reality which are not directly related to feeding, reproduction, and other activities that have been selected through biological evolution for their survival value.

There are numerous examples of how our limited abilities to perceive and process reality affect our everyday life. Playing the lottery, for example, can not be justified by any rational argument; however, a large number of people engage in gaming and other irrational activities. They play with odds that are clearly against them.

Another example is laughing, frowning, or scowling when speaking over the telephone. Our face is not perceived by our interlocutor, but our speech is generally not dissociated from our emotions or facial expressions.

Modern economics is especially aware of irrational human behavior. Herbert Simon, Daniel Kahneman, Ernst Fehr, and many others have dedicated their lives to studying and understanding several irrational economic behaviors of humans.

A famous limitation of human thought is the learning curve. That is, it takes time to get familiar with the challenges of new situations. It implies that there also exists a forgetting curve. We have to discard what we think we know. The deeper we dig into a field of knowledge, the more difficult it is. A classical psychological term called "inattentional blindness" illustrates this point. It comes from an experiment in which people were asked to watch a video of a basketball game and count the number of passes. Halfway through the video, a gorilla walked through one door of the arena and out the other, but half of the people didn't see it. When you tell people to focus on one thing, they frequently do not recall other things. This classical psychological experiment illustrates the point that when people acquire a certain amount of knowledge, they become blind to novel views and facts since everything is filtered through their expertise. Management gurus call this "educated incapacity." The more you know, the harder it is to see anything new. Thus, expertise and specialization limits the discovery of new features and novel laws of nature.

PHILOSOPHY OF SCIENCE

"What we find out in philosophy is trivial; it does not teach us new facts, only science does that. But the proper synopsis of these trivialities is enormously difficult, and has immense importance. Philosophy is in fact the synopsis of trivialities." . . . *"In philosophy, we are not, like the scientist, building a house. Nor are we even laying the foundations of a house. We are merely 'tidying up a room'."*

—Ludwig Wittgenstein 1930

It might be significantly more efficient if the person that is tidying up the room is also the same person who actually lives in the room. Although external help is always welcome, tidying up a complex multidimensional room is no easy task.

As mentioned in the introduction, the philosophy of science, the analysis of scientific phenomena viewed through the lens of physical science, has produced important insights into the workings of science. We summarize the most important contribution to this understanding of scientific progress by the insights that scientific progress is punctuated rather than continuous (Thomas Kuhn: *The structure of scientific revolutions*); that disproving a scientific theory is much more important for scientific progress than proving it (Karl Popper: *Logik der Forschung*); and that scientific dynamics can not be planned ahead or controlled from the outside (Paul Feyerabend: *Against Method*). These deep insights are relevant for all disciplines of science and are part of our evolving modern view of science. Curiously, these three insights are analogous to the driving principles of biological evolution: evolution suffers catastrophes and is characterized by punctuation followed by bursts of diversification (Stephen Jay Gould); evolution is driven by natural selection which weeds out unsuccessful forms or solutions of life (Charles Darwin); and random events and the stochastic nature of mutations are the engines of evolutionary progress (Gregor Mendel).

Attempts by philosophers, such as Georg Wilhelm Friedrich Hegel, Immanuel Kant, Francis Bacon, George Berkeley, Bertrand Russell, and Ludwig Wittgenstein, to understand science have successfully molded the *Zeitgeist* and the human outlook on the world, but have had little impact on the instrumental implementation of science. They were based on pure brain power rather than on application of the scientific method. Thus, they failed to understand the underlying heuristics of science. It seems that only a Texas cowboy can understand the Wild West and only a true sailor can understand the diverse challenges of sailing the ocean. Yet, experience has not been a limitation for humanity. A lack of knowledge of Texas does not stop people with absolutely no cowboy experience from writing about cowboys or writers that have never put a foot on a boat from describing the dangers of sailing on the high seas. Fantasy is important for humans and has a role in any heuristic epistemology, as it allows for the construction of complex mental models of humans and our interactions with the world. Importantly, fantasy without other mental aids is limited. This same reasoning should be applied with even more logical force to complex human endeavors, such as the working of science. It is unlikely that someone who has not spent days, nights, weeks, or years attempting to employ nature to answer questions, by following strict scientific guidelines, will have a deep understanding of the workings of science. Thus, it is unlikely that philosophy, without the input of experimental science, will allow for a deep understanding of science.

Sociologists, philosophers, and other students of the scientific process often write about science literature and not directly about science and its methods. Surprisingly, very little has actually been written about the scientific method. It seems that most scholars assume that science and its method are closed and finished subjects that does not need any further attention.

A good way to understand science is through its history. Many histories of science have been written after Galileo's time (see, for example, Thomas Kuhn, Karl Popper, Joseph Shumpeter, Paul Feyerabend, and Imre Lakatos). I do not want to explore them here as the reader is better advised to look for works by these authors and other extensive analyses of the history of science. Here, I want to understand science from the standpoint of modern scientists. I want the reader to look at science from the laboratory bench, and not from the heights of sophisticated intellectual constructs.

PARADIGM SHIFTS

Ideas are certainly important in guiding human thought and action. Ideas persist for a long time and old ideas are still among us. The XIX century was strongly influenced by ideas first expressed in the XVIII century. Ideas from David Hume, Voltaire, Adam Smith, and Immanuel Kant marked the liberalism of the XIX century. Similarly, ideas by Georg Wilhelm Hegel, Auguste Comte, Ludwig Feuerbach, and Karl Marx influenced the rise of totalitarianism in the XX Century. Other ideas that molded our culture were of a more fundamental kind, and thus, were slower to permeate human thought. The discovery by Galileo of the limitations of human senses and the need for experiments to evade the tricks and limits imposed by our minds on our understanding of the phenomena surrounding us has still not permeated the minds of the gentry. Nor has insight into biological evolution, gained through the intellectual efforts of Jean-Baptist Lamark, Alfred Russel Wallace, and Charles Darwin (and through the experimental evidence of geniuses such as Leonardo da Vinci), molded the thoughts of the common and not so common citizen.

The workings of science have been studied mostly by philosophers and sociologists, and more recently by anthropologists. Scientists claim that science can not be learned through textbooks, but requires extensive training in the laboratory or in the field for practitioners to become even faintly acquainted with the scientific method. These differences in outlook provide different conclusions. It may be time to attempt a more scientific analysis of science.

Let us analyze science using its own method. Let us imagine that humans were studied by a Martian biologist. He would probably question the adaptive value of science for humans, asking if science is a miscellaneous event in history or a necessary step in biological evolution. He might try to research whether science increases the biological fitness of the individual. If he finds a positive correlation, he might ask if this increase in fitness is achieved through selective forces acting on the individual or on the group. That is precisely what we want to answer next.

Biological fitness, of course, can be measured in many different ways. For an evolutionary biologist, two components are probably the most important: longevity and fertility. Fertility in an ecological situation of overpopulation is of little value. It will not correctly estimate fitness in societies with a large variance in survival probability. Thus, longevity seems to be the best indicator for fitness in modern human populations. The following graph shows how longevity, measured as the life expectancy of children born in 2000–2005 (according to the Human Development Report of the United Nations, published in 2005), of a given country is correlated with the scientific activity of that country (measured by the number of publications reported by SCI -Thomson Scientific, per capita, for 2003). We observe that the countries with the highest life expectancy also have the highest scientific activity. The more science that a country supports, the longer its citizens live, although very small increases in the average longevity of a nation correlate with large increases in

scientific development. We can then affirm that in societies with highly developed scientific activity citizens live longer. Scientific activity correlates
exponentially with longevity, the most relevant component of fitness for modern humans (Figure 1.9).

Certainly, many other factors in a modern human society correlate with human fitness. The classical correlate, of course, is with average individual
wealth. The degree of scientific development in a society correlates strongly
with average income per capita, as do other intellectual and epistemological
human activities, such as research in social sciences and humanities, production of films, general education, etc. All of these features could be assumed
to correlate with proxies for human fitness. Of these features, science shows
the strongest correlation with average individual wealth.

In Table 1.1, I present the correlation coefficients between per capita indicators for economic wealth provided by the World Bank. The indicators are:
Gross Domestic Product per capita in Purchasing Power Parity (GDP), Human Development Index, Investment in Research and Development as % of
GDP, and per capita indices for scientific and artistic productivity for 44 selected countries. The indices for scientific productivity are: number of publi-

**Figure 1.9. Plot showing the relationship between the numbers of scientific articles
published in 2003, per capita, of a country (SCI) and the average life expectancy for
2000-2005 of its inhabitants. Data from UNDP.**

Table 1.1. Correlations between selected country variables (Marked correlation coefficients are significant at the level p<0.01)

Data for 2003	Gross Domestic Product	Human Development Index	Investment in Research and Development
Publications in PubMed database per capita	0.93	0.64	0.61
Science Citation Index publications per capita	0.93	0.69	0.66
Social Science Index publications per capita	0.61	0.42	0.49
Arts and Humanities Index publications per capita	0.61	0.40	0.49
Movies per capital	0.73	0.59	0.51

cations each country produces in biological and medical sciences that are recorded in the PubMed data base run by the National Library of Medicine in the USA; the number of scientific articles that researchers in each country publish, as determined by Thomson ISI in natural sciences, social sciences, arts and humanities, and the number of movies of all types produced in the country, as recorded by the International Movie Data Base (Movies). These numbers were collected for the year 2003 and were divided by the population of that country to obtain per capita indices. The data shows that rich counties produce more science, indicating that science and wealth are highly correlated (correlation index of 0.93). All other variables showed lower correlation coefficients with GDP/capita, indicating that they relate less strongly with wealth compared to science.

In addition to affecting our social fitness (GDP, life expectancy, etc.), science is at the heart of our modern technological society. What makes science so much more adaptive than other heuristics or epistemological human constructs? The next section addresses this question using an analytical approach based mainly on evolutionary thinking.

Chapter Two

The Ascent of Empirical Science

When thinking changes your mind, that's philosophy.
When God changes your mind, that's faith.
When facts change your mind, that's science.

—The Edge Annual Question, 2008

THE ROOTS OF LOGIC

I invite you to visualize a more reduced time window to view the history of the human race, spanning the last 3,000 years. An important aspect of this time window is that the ideas and mental models of humans can be read directly from the scripts that they produced. We can understand our antecessors' ideas without needing to speculate about what they wanted to say or what they thought. We have their thoughts, at least part of them, in our hands and can read directly from their writings.

The classic Greek savants had many features that are the envy of most modern scientists and that might be considered as characteristics of a scientist. These attributes include critical thinking, effort placed on accurate descriptions, rational analysis of facts and ideas, logical thoughts, extreme ingeniousness, and intelligence. Most Greek savants did not produce methodological algorithms that could be called experimental science, although some were very close to this achievement. What was missing?

The answer to this question is fundamental to the understanding of the nature of modern science. Before we address this question, a number of additional relevant concepts should be introduced.

Figure 2.1. Adapted from the color painting "Ingravity Phenomena" by Remedios Varo.

INDUCTION, DEDUCTION, AND RATIONAL THOUGHT

The two methods of thought most widely promoted are undoubtedly induction and deduction.

For David Hume (1711–1776), Max Born (1882–1970), Karl Popper (1902–1994), and many others, induction, although fundamental for science, has its limits. Born wrote that.".. no observation or experiment, however extended, can give more than a finite number of repetitions; therefore, 'the statement of a law—B depends on A—always transcends experience." This kind of inductive reasoning is essential for rational thought and serves as the basis of science.

Deduction, on the other hand, is a sequence of statements, such that every statement can be derived from the preceding statement. Based upon one observation or one statement, the rational mind, using deduction, can build a series of rational statements. The limit of deduction, of course, is that it leaves open the question of how we prove the first sentence to be right. No amount of deduction can root reasoning on objective first principles.

These rational processes leading from the general to the specific or from the specific to the general are the basis of all reasoning. Even animals show rudiments of this kind of reasoning. For example, deduction, in the language of the predator-prey game, can be exemplified by a fast approaching form that gets bigger faster, which allows the observer to deduce that the approaching object is a large organism that might try to harm or eat them. The observer deduces that a fast approaching object may be a predator, and thus should be avoided. On the other hand, objects that get smaller are probably prey that try to escape the area of influence. Applying induction allows us to extrapolate from past experience, as large organisms are normally stronger than we are and frequently try to kill us. Large objects could be potential predators and it is advisable to avoid them. Thus, both induction and deduction simplify the relationships between objects that we have to handle and help us to develop abstractions of reality.

Both forms of reasoning are based upon the presumption, common from Aristotle through the medieval scholars and Thomas Aquinas to many modern intellectuals, that our mind is a fountain of truth that predates experience, and that our intelligence makes us capable of understanding terrestrial and supra-terrestrial phenomena, from the workings of nature to the existence of God. This kind of attitude is often called philosophical rationalism and is based upon the belief that human reason is the only legitimate fountain of true knowledge.

For our mind to perform useful rational thinking, it must classify the objects of study. Once a large amount of information has been transmitted to our brains, we need to classify the objects according to properties that are easily detectable and recognizable in order to efficiently apply deductive or inductive reasoning. Classifying objects that move away from us as prey and objects that move towards us as predators, simplifies our world view enormously, allowing us to develop behavioral responses which can be triggered quickly and efficiently. In some cases, our simplifications may lead us to make mistakes, as it is only a crude approximation of reality or an abstraction. Rationally organizing shared features and identifying characters that are specific to a given object is fundamental for initiating a systematic study of any part of nature. This activity of classification of organisms into specific taxa, which is performed systematically and methodologically, is called taxonomy.

Knowledge is often accumulated fortuitously. Rationalizations of reality be it taxonomic classification of objects, inductive or deductive reasoning, or modeling reality and its interactions, do not always allow for new insights into reality. Only if we have precise questions to ask of nature, might rational reasoning be helpful. If, for example, we want to study a pendulum, we need to know first what questions we want to answer by studying the pendulum. Given the multidimensional nature of reality, we might be interested in analyzing the metal alloy that is used to build the pendulum; we might be interested in the aerodynamic form of the weight that reduces friction with the air; or we might be interested in the dynamics of the movement that produces regular oscillations. Thus, our interest or motivation will define the question that is asked, which in turn will start a route of scientific exploration. For each specific route different rational dimension will be explored and different types of classifications and reasoning will be used.

Knowledge is not always advanced by rational reasoning since reality is not always amenable to rational analysis. Psychoanalytic analysis, political decision making, and even simple socializing at a party do not improve with rational reasoning. Practice, intuition, and other epistemological solutions seem to be more appropriate than rational science in these cases.

LEVELS OF THE KNOWABLE

Our relationship with the world around us may become very complicated. Some phenomena and some objects can be directly perceived by our senses, while others can not. The human brain is trained to understand and follow a relationship of networks, such as mechanical cause-effect chains, whereas we can not comprehend other relationship networks, such as stochastic cascades, compounded risks, non-linear webs, and multidimensional phenomena. The psychoanalyst Ricardo Palma proposed a graphical representation of the domains that relate the mind to our surrounding world (Figure 2.2). Each sphere or domain requires different mental tools. Instinct, intuition, and a large dose of self-esteem help us to deal with the unknown but knowable. Verbal and literary skills allow us to tackle the describable with new words and metaphors. Rational scientific thinking allows us to explore that which can be measured and predicted.

The transitions between two consecutive spheres provide us with insight into the heuristic advances of human epistemology. These transitions might correlate with fundamental features of our thinking. The transition from the unknowable to the knowable, for example, is related to "consciousness," i.e., the ability to perceive or know of the existence of something.

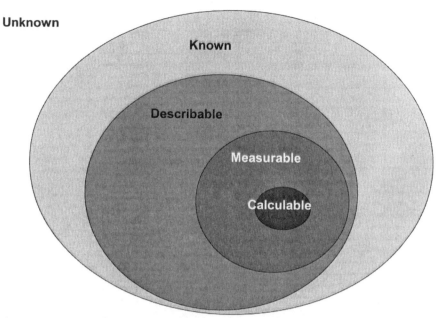

Figure 2.2.   Adapted from the psychoanalytical theory of Ricardo Palma.

The transition from the knowable to the describable requires a model, even a crude one, of the known object. The description of something implies the assignment of properties, the use of words that relate the object to other known objects through the discovery of similitudes or differences. This transition implies an exercise in classification whereby the object is put into context with regard to other objects. Classification allows objects, phenomena, and concepts to be placed into abstract categories which allow better descriptions, thus a more detailed classification is related to a finer description.

The transition from describing an object to measuring some of its properties implies a fundamental heuristic step towards a more objective (possibly quantitative) description of the object. This step allows for a more scientific classification and for replication of observations that are amenable to scientific experimentation and explorations.

The transition from the measurable to the predictable implies the construction of detailed predictive models based upon falsifiable theories. These theories can be expressed mathematically or numerically and are at the center of modern scientific endeavors.

New analytical tools, theories, and devices to detect signals that are hidden from our senses, might expand the spheres of the measurable and the predictable at the expense of the spheres of the describable, the knowable, and even the unknowable, represented in the Figure 2.2. That seems to be the responsibility of future scientific developments. How do we deal with social and psychological phenomena that are not completely understood, nor even detected and appreciated to their true extent and importance? This is certainly one of the most challenging questions of modern science and some attempts to resolve it are explored here. Our recent past history and our positivist optimism allows us to believe that the future will provide science with extraordinary success in shrinking the spheres of the unknowable and expanding those of the predictable.

Kurt Gödel (1906–1978) analyzed an analogous problem in purely mathematical terms in his theorem of incompleteness published in 1931. This theorem states that in any recursive axiomatic auto-consistent system that can be described by with natural numbers you will find true propositions that can not be proven from its axioms. That is, you need axioms from outside the system. If we relate this proposition with those referring to the limitations of human thought by Kant and extrapolate its way of thought, we arrive to the proposition that in order to describe exhaustively and completely any given system, we require a system that is nearly as large or larger than the one we want to describe. In the words of Jorge Luis Borges (1899–1986) .".*.In this beautiful empire, the art of cartography reached astonishing perfection, so that the map of a single province occupied a whole city and the map of the empire a whole province . . .*"

According to this logic, a system much smaller than the universe, such as our brain, will never be able to grasp the complexity and the immensity of the cosmos. In terms of our graph, the unknowable will always be part of life. Modern science, as ancient science did in its time, will have to find ways to cohabitate harmoniously with psychological and social constructs that humans have developed and use to deal with the unknown.

DEFINING SCIENCE

Science can be defined in many different ways. One view involves a collective effort of a multitude of pragmatic researchers who base their conclusions on the careful and progressive research of facts, with great respect for objective evidence through the constant and systematic use of experiments. Scientists wait before formulating grand theories until the strength of the empirical facts make it unavoidable, discarding the sole use of intuition, except for the initial formulation of new theories. Although science necessitates the use of rational constructs, it advances through scientific observations and experiments.

Several qualities are associated with science: pulchritude, precision, rigor, cleanliness, thoroughness, logic, intelligence, resourcefulness, attention to detail, synthesis, creativity, beauty, simplicity, complexity, quantification, etc. Yet, all of these qualities can be assigned to successful practitioners of many disciplines; they do not seem to be exclusive to science.

So, how can we define science? What are the exclusive qualities of science? What differs between the Galilean revolution and the classic Greeks?

We can define at least three factors that are basic to modern experimental science:

a- Scientific theories must be rational and logical, so that any instructed human being or computer can understand them.

b- The human mind is limited as its evolution was driven by biological forces to produce and guide behaviors aimed at gathering resources, hunting for food, defending against predators, reproducing, and socializing. The human mind was not evolved to produce science. The human mind suffers many and deep limitations regarding its ability to rationally understand the world. Based solely on the production of rational constructs, without further aid, the human brain is unable to grasp the details of the surrounding physical, chemical, biological, and social world.

c- Scientific theories have to be refutable or falsifiable by experiments; they must allow for experiments that can refute them. That is, a theory, in order to be considered scientific, has to allow for the possibility that experimental evidence might prove the theory to be false. Proving a theory right is not enough, it needs to be phrased so that it can eventually be shown to be wrong experimentally, if it is incorrect. The experiment, the empirical observations and manifestations of nature, have to prevail over any product of our limited mind. Science recognizes that reality overwhelms our imagination. This statement indicates that science is a process where no definitive truth exists. Science is based on hypotheses that can be falsified and replaced by even better, more general, and more detailed hypotheses.

These three components are indispensable for science. Each of them can take multiple forms. The construction of rational theories can become highly sophisticated and is sometimes the only aspect with which a given academic discipline is interested (philosophy, mathematics, logic). Experimenting with reality and falsification of scientific theories can become tricky and indirect, as is often the case when studying complex systems or when exploring metascience. Our mental limitations are also changing, as they depend, in part, on our education, the way we perceive the world and how we communicate with each other. Our limitations can be overcome by the extensive use of technology. This affects the way in which we interact with and perceive our environment.

Even though other features might be important, such as creativity and imagination, science can be nurtured and prosper without these features and, therefore, they are not necessarily always part of science. For example, computers often design and aid in the production of novel compounds and drugs in the laboratories of pharmaceutical companies; however, the concept creative or imaginative are not applicable to computers.

Science shares many additional attributes with other intellectual human activities. Science requires rational thought, logic, the construction of concepts and models that establish relations among them, communication skills, technical abilities, intelligence and insights, originality, creativity, rigor, perseverance, hard work, discipline, tradition, social stimulation, and many other features. Often, these attributes can be developed extremely successfully by shadowing other persons that possess the chosen skills. But it is the presence of the three characteristics mentioned above, regardless of the mastery of other skills, which allows us to recognize an activity as scientific.

Often, a good picture is worth a thousand words. This is the case of the drawing by Rob Gonsalves, shown at the beginning of this chapter, which is reminiscent of the metaphor by Plato which describes chained men in a cave, looking at the shadows of the external world. They acclimated to viewing reality through these shadows such that when one of them saw the real world and relayed his experience to his companions, they assumed that he was crazy. Science is the only known and tested heuristic human construct that allows us to advance our knowledge of measurable reality. Science does not possess truth. It works by building hypotheses, using rigorous methods and sophisticated instruments to refine observations and design experiments, discarding unsuccessful or defective hypothesis, and constructing new ones. Science works by building on humility (the acknowledgement that we do not know everything) and sustained effort.

As science seems to depend on humility and accepting experiments as the ultimate judge, let us dwell on these aspects of human consciousness.

ARROGANCE MAY LEAD TO IGNORANCE

If our brain is just a small faulty subset of the world, it will never be able to grasp the complexity and immensity of the cosmos. A series of recent Nobel laureates in economics demonstrated the irrationality of several human economic behaviors and sensory psychologists have mountains of evidence that the human mind is easily tricked. Everyday life provides a multitude of examples of how incapable the human mind is of correctly assessing risk and probabilities, giving rise to a whole industry of lotteries, money games, betting, and other irrational human activities.

One of the first tasks that the human mind must solve is to understand the neural signals that are produced by our sensory organs and sent to our brain. The innumerable signals are filtered and processed to produce an abstraction of reality. The meaningful stimuli are translated so we develop behaviors that are compatible with the surrounding world. When, for example, visual signals are filtered and processed by our retina and brain to increase contrast and enhance the perception of forms, we can detect other organisms with greater precision, so that we can either approach them if we perceive them as prey, or avoid them if we perceive them as predators. This process of filtering, which is indispensable for survival in the real world, leads to errors, lack of awareness, and mental limitations.

Despite the mountain of evidence regarding the limits of the human mind, most humans believe that rational thinking, using only the powers of our mind, is sufficient to untangle any complexity in our surrounding reality. A strong belief in the absolute power of our rationality is irrational.

Some humans, past and present, have recognized the limits of the individual mind, accepting it as a fact of life. They place their faith in the hands of religion or dogma. Although this attitude may have several advantages, it does not allow us to explore the underlying mechanisms that limit it. Although this faith in a superhuman guidance is very common today and was accepted by most medieval scholars and classic Persian, Egyptian, and Greek savants, it can not be considered as a rational solution to the limits of the human mind. .

An interesting explanation for our mental limitation is given by evolutionary biology (Konrad Lorenz: *Die Rücksite des Spiegels*). Although many mental algorithms, natural logic, and mental quirks can be explained by their adaptive advantages in the struggle for survival in the practical world, they do not help us to unravel the deeper truth underlying reality. Thus, the human mind has adapted to find *satisficing* solutions (Herbert Simon 1916–2001); that is, to produce models that allow us to survive and reproduce successfully, but not to untangle the mysteries of the sub-atomic world, or those of the cosmos, or those of complex dynamic systems.

One important consequence of accepting our limitations is the need to promote tolerance (Baruch Spinoza 1632–1677). A measure of degrees of tolerance should therefore serve as an indirect estimate of the levels of acceptance or recognition of our own limitations. Such measures of tolerance are available. The Pew Global Attitudes Project explored the prevalence of specific human attitudes in a number of countries. Some of these attitudes are related to tolerance. Table 2.1 shows correlations between positive answers to some

Table 2.1. Correlation coefficients between answers of a World-wide Pew survey and indices for economic, cultural and scientific development

Pew's attitude 2002 vs. ICI 2003	GDP	HDI	PM	SCI	Movies
Homosexuality should be accepted by society (N=41)	0.64 ***	0.74 ***	0.56 ***	0.66 ***	0.61
It is not necessary to believe in God to be moral (N=38)	0.68 ***	0.73 ***	0.62 ***	0.79 ***	0.61
Success is not determined by forces outside our control (N=44)	0.44 **	0.34 ***	0.44 ***	0.31 ***	0.52
Religion is a personal matter and should be kept separate from government (N=39)	0.33 *	0.11 *	0.31 *	0.38	0.29

***, **, and * indicate values that are statistically significant at the probabilistic level of p < 0.001, 0.01, and
   0.05 respectively
GDP = wealth of the nation given as Gross Domestic Product
HDI = well-being of the nation's citizens given by the Human Development Index
PM = Scientific productivity as measured by PubMed of the National Institute of Health in the USA
SCI = Scientific productivity as measured by the privately run Thomson's Science Citation Index
Movies = Number of movies produced annually
A correlation of 1 indicates a total correspondence between the two variables that are being compared. A
   value of 0 indicates no correspondence at all between both variables.
Data from *Science, religion, and economic development*, Interciencia 2005.

relevant questions from the Pew Global Attitudes Project and indicators for economic development and scientific productivity. From this data, it is very clear that countries with high scientific and economic development had more tolerant citizens with respect to sexual attitudes and religion.

A similar survey of the attitudes and values of a group of scientists, previously classified according to their scientific productivity (number of papers) and the impact of their scientific work in the scientific community (number of citations), showed that the most productive and successful scientist generally had the most modest attitudes regarding their awareness of their limited impact on society (Interciencia 2005). This correlation was independent of their reported self-esteem, which was relatively very high. That is, some forms of humility correlated with scientific success. These relationships are explored more extensively in a different context by the economist Richard Florida in his book, *The Flight of the Creative Class*, where he produces a global competitiveness index of nations in terms of the "3 Ts" of economic growth: technology, talent, and tolerance. Again, tolerance was found to be essential for economic growth and correlated with technology.

The recognition that rational thought can not be the final arbiter of truth and that some mechanisms that are external to the working of our minds are essential in the pursuit of new knowledge, is fundamental to modern experimental science. This insight differentiates science from philosophy, dogma and religion and leads to the acceptance of observations and experiments as the ultimate arbiters of a rational dispute and as the source of new insights into the world.

The basic humility that serves as the foundation for science leads to the acceptance that no scientific theory is absolute truth; rather it is an approximation of it. Any scientific theory, regardless of the authorities that support it, might be shown to be false by empirical facts.

THE EXPERIMENT AND SCIENTIFIC OBSERVATION

Science is based on logic, rational thinking, models, technology, intelligence, hard work, group synergy, sustained effort, past experiences, etc. Many of these features are shared by other human activities. Furthermore, many academic disciplines and intellectual endeavors have developed skills and tools far superior to those that most scientists can attain. Artists, musicians, writers, and many intellectuals share most or all of these features and are much better than most scientists at communicating subtle ideas and feelings to fellow humans. Humanists and philosophers are often far superior at developing and explaining long chains of logic and complex rational arguments than scientists. Literati and poets excel in combining different aspects of reality and drawing associations between distant facts.

The use of experimental evidence as the last arbiter of any dispute is the main characteristic of science. It is also the motor that advances science. Reason and logic are part of the scientific method, but, due to the acceptance (consciously or otherwise) by scientists of the limitation of the human mind, the experiment outwits reason.

This curious relationship between reason and experiment leads to a unique definition of objectivity in science. It is not so much the distance between the subject and the object of study that defines the degree of objectivity, although a larger distance certainly helps objectivity; it is falsifiability that operationally defines objectivity.

Falsifiability was defined by Karl Popper in the 1930's as an apparently paradoxical idea: a proposition or theory cannot be scientific if it does not admit the possibility of being shown false. For a proposition to be falsifiable, it must be at least in principle possible to make an observation that could show the proposition to be false. For example, the proposition "all swans are white" would be falsified by observing one black swan. The discovery of a black swan would show the hypothesis to be false, forcing us to limit the hypothesis to a narrower range: "All swans in Stockholm are white" or use a wider range: "Most swans are white."

Falsifiable means that it is possible to prove it wrong. If, when we test a hypothesis, we perform the test in such a way that a priori all possible outcomes will confirming the hypothesis, then we have not performed an experiment at all. We need tests that might prove the theory wrong.

For example, a researcher of extrasensory perception (ESP) frequently finds a number of individuals who seem to be able to read minds. Typically, a person thinks of any number between 1 and 1,000 and writes it down. Next, the person is asked to sit next to the ESP subject and think of the number. The ESP subject is told to guess the number. The statistical odds are that we might find a number of people who can guess it more often than expected by chance. A test aiming to falsify ESP will repeat the experiment with different persons and calculate if the ESP subjects will guess the correct number more often than chance. A non-falsifiable theory, however, such as ESP, claims that ESP abilities are fragile and likely exist only under optimal testing conditions, so that repeated test should not be performed. This theory can not be shown to be false.

Producing falsifiable theories is not an easy task and the degree of falsifiability of a theory is its strength. Good theories are easily falsifiable. A person who could show that objects on the earth's surface levitate would prove the theory of gravity false. Thus, the theory of gravity is very easy to disprove, but nobody today has been able to do so, demonstrating that the theory has a strong scientific basis.

"We are all capable of believing things which we know to be untrue, and then, when we are finally proved wrong, impudently twisting the facts so as to show that we were right. Intellectually, it is possible to carry on this process for an indefinite time: the only check on it is that sooner or later a false belief bumps up against solid reality, usually on a battlefield." as stated in George Orwell's 1946 essay "In Front of Your Nose." Falsifiability must avoid *ad hoc*, *post factum* amendments to the theory that are adapted to the outcome of experiments. Once a theory has been shown false, all options should be reconsidered and a new hypothesis or theory should be built. This new theory, of course, will have to draw on former falsified theories, since disproving a complex theory does not necessarily invalidate all aspects of that theory. This way of advancing thought is radically different from the more traditional construction of logic chains of ideas. For science, the logical consistency of a theory, although important, is second to the importance of consistency with empirical facts. Many scientific theories have very irrational and illogical aspects. One such case is that of quantum theory. However, it is the falsifiability and the outcome of innumerable experiments, not the rational beauty, that make the theory scientific.

Given these criteria, can we regard mathematics as a science? If mathematics deals with the study of logic, which is a product of our mind, it would be justified to dismiss it as non-scientific. But mathematics also deals with the underlying logic of features that are outside of our mind, such as relationships between physical objects. Our criteria allow for a precise answer to this question: if a mathematical theory is falsifiable, it can safely be classified as scientific.

The need for empirical verification of theories leads a scientist to be sparse in constructing long and complex chains of rational thought. In contrast, philosophers excel in constructing castles of rational thought without regard to its empirical falsification. Scientists worry that each new thought or idea, added to an existing scientific theory, might be false, and needs empirical evidence to justify its inclusion. Therefore, the habit of producing long rational constructs without an empirical basis for each part, is not scientific and strongly differentiates scientists from philosophers and other non-scientific intellectuals. Experimental scientists know that the probability that any new idea is false is very high. Therefore, the probability that a chain or series of non-tested ideas correctly reflects a new unknown reality is minimal. Recognition of the limits of rational thinking should make the enlightened intellectuals more cautious when producing sophisticated political discourse, long philosophical discussions, or purely verbal arguments in general. This way of thought, if applied systematically, prevent us from traveling down the long, costly, blind alleys of unfruitful intellectual research. This skeptical attitude accelerates the acquisition of knowledge and allowed the fantastic technological developments of the past two centuries.

TECHNOLOGY

The majority of humans come in touch with science through technology, and thus confounds both. Science and technology are sisters with different aims. In science two activities are essential: observing nature, and performing experiments that ask nature specific questions. Both of these activities are greatly enhanced by devices, apparatuses, and machines that extend our senses and increase the power of our biologically inherited organs. These extensions are achieved by technology. Although technology grows on science, it aims to solve problems. Understanding nature is important only if it helps solving applied problems.

Technology and science are not the same but they need each other. Science aims to understand reality, while technology transforms reality. For technology, a coherent complete explanation of reality is not required; rather, any explanation that produces practical results in the chosen domain is welcome. Science, on the other hand, ignores applications to improve human comfort when exploring reality. It follows the intricate fractal-like ramifications of a multidimensional reality wherever it might lead, whereas research in technology is focused on a specified outcome.

Technology and science often use similar research methods; however, the emphasis on the explanatory power and use of a given theory differs. Neither discipline can exist without the other. Technology is based on science. Technology builds on scientific laws to develop instruments and devices that in turn help scientists explore nature more accurately and develop better scientific laws. These results then allow for technology to produce more sophisticated and powerful tools and devices. Modern technology can not exist without science and modern science can not exist without technology. This symbiosis is the backbone of modern technocratic societies.

ELEMENTS OF SCIENCE: OBSERVATION, DESCRIPTIONS, MODELING, AND COMMUNICATION

Besides recognizing the limits of our mind and accepting experimentation as the ultimate arbiter, science draws on many other human abilities. Clear thinking, logical thought, abstract reflection, precise observations, and sophisticated communication are fundamental to science. This fact, plus the ever increasing specialization and compartmentalization of human activity, leads many scientists to specialize in specific activities that, although part of a scientific project, do not involve rigorous application of the scientific method as described here. Rather, these activities rely on other disciplines to complete their tasks.

For example, many taxonomists engage in describing new species using tools and methods that do not promote falsifiability. They strongly draw on their own intuition to define a new species based on the available dead specimens at hand. Most taxonomists have no means of experimentally testing whether a newly described species is reproductively isolated from its sibling species. Nor can they be sure if a single specimen collected in an inaccessible place is a mutant, a defective formed individual, or a true new species. Geneticists, natural historians, or ecologists working in the field may eventually be able to falsify the status of a newly described species. But, even if a taxonomist produces a classification scheme that is not easily falsifiable, their work is fundamental to evolutionary biology and ecology, which are disciplines that rely more strongly on experimental verification than taxonomy.

Other examples might be found in theoretical computer science, which includes the study of artificial life, cybernetics, and artificial intelligence. The simulation models used by these scientists are often inspired by non-falsifiable philosophy, subjective logic, and non testable sociological theories. The scientific value of the derived algorithms that simulate features in complex systems and societies, however, are tested by experimental economists, field biologists, and anthropologists, who design experiments and observations in an attempt to disprove or support the conclusions of the simulations.

Communication is becoming one of the most important abilities of successful scientists. Publish or perish is the rule among modern scientists. Having good ideas or novel experimental results are not enough to guarantee publications. In an ever more complex and bigger world, communicating complex ideas simply and placing isolated experimental results in an attractive context is fundamental for the professional survival of a scientist. These skills, however, are generally not associated with science. Academic curriculum largely ignores these skills despite the fact that journal editors and the scientific public at large value good communication of scientific data and ideas. The future of a successful scientist will be more related to good communication and communication skills will become ever more important for science.

Many skills and intellectual abilities are valued in scientists. Logic rigor, abstract reasoning, model building, summarizing large bodies of knowledge, cooperating in interdisciplinary groups, creating and accepting trust, reliability, motivational capabilities, precision, attention to detail, hard work, intelligence, social skills, along with many other factors are fundamental to good and efficient scientific activity. These attributes are important for many human activities, whether they are related to science or not. Since these activities are not exclusive to science, they are not discussed here. This does not mean that they are unimportant or that students trying to achieve success in science should ignore them.

LIMITS OF SCIENTISTS, SCIENCE, AND ITS APPLICATIONS

All human endeavors have limitations. After all, humans are not Gods. Here, I name just a few of the most common limitations that are experienced by scientists and science:

Human weakness: Science, after all, is performed by humans and thus suffers from all human limitations. Greed, ambition, politics, dishonesty, shortsightedness, prejudice, etc. can not be excluded from the working environment of scientists, but they clearly do not form part of science or its method.

Studying the measurable: Scientists can study the measurable and are very good at handling the predictable (see Levels of the Knowledgeable, above); however, everyday life is full of phenomena that are not measurable and/or predictable. Thus, these events can not be tackled by science, at least for the moment.

Science is useful for discovering knowledge, not for applying it: Many human activities involve assertive decision making or decision making using few resources and limited information. In this case, it is better to use experience, thoroughly tested engineering tools, or widely used algorithms, for action rather than applying the scientific method. Science is good for the advancement of knowledge, but not for its application. For example, physics advances our knowledge of the world that we touch and see, but it is engineering that builds upon the knowledge that is unveiled by physics. In this same streak, biology, physiology, and life science uncovers the mysteries of life while medicine, agricultural engineering and veterinary science apply much of this knowledge to solve specific problems. This difference does not mean that a single person can not excel in both disciplines; rather it suggests that not all scientists are good engineers nor are all engineers good scientists.

Pure science differs from applied science: Many scientific disciplines, however, have not separated application from the discovery of laws and facts. The clear separation between physics and engineering is absent in most academic disciplines. For example, economists and sociologists mix projective wishful norms with the study of actual real phenomena. That is, the desire to build a specific society or to achieve a specific outcome influences rational thinking in the analytical phase of these disciplines, since it has an effect on objective thinking similar to that of dogmas. The drive for applications hinders free thought and limits explorations of novel routes.

Scientific explanations do not necessarily satisfy psychological needs and anxieties, the need for affection, the desire for success or power, or many other human motivators. Science is useful for discovering the unknown, not for satisfying human needs and wants.

The pace of innovation is slowing: Several studies show convincingly that the pace of scientific innovation is slowing today. The emergence or boom of ideas, cultural traits and technologies, and their later dissipation from human interest, is a known and proven fact of human history. Can it be that science will loose its grip on the human mind and fade away, being substituted by other more "post-modern" ideas? Or is reality becoming ever more complex, escaping the analysis of even the most sophisticated science? Have we already obtained most of the knowledge that science can harvest from nature? We can not negate these questions with certainty but we might envision a future where more interdisciplinary endeavors will expand the applicability of science to new fields and more complex science will emerge. This might put new constraints on scientific activity and may, in the end, be limited by the capacity of humans to tackle complex problems.

> *Il faut contempler; il faut penser. Qui pense peu se trompe beaucoup*
> *"We need to observe; we need to think. Who thinks may commit many errors"*

> —Leonardo de Vinci

Chapter Three

Modern Science

The smallest, contains the infinite

— Tao

ON THE MULTIDIMENSIONALITY OF REALITY

The structures of reality and the world have curious dimensions. Our mind simplifies the world by intuitively projecting every object we see into four dimensions: three spatial dimensions and one temporal dimension. But, when we simultaneously tackle webs of objects and their interactions, this simplified four-dimensional world does not provide a sufficiently precise description that we can grasp mentally. New conceptual dimensions are necessary for this description.

The multidimensional properties of the world can not be collapsed into the multi-dimensional concepts of physics. A world of various dimensions as envisioned by physics, which includes length, width, height, and time, contemplates a continuous plane intersecting another continuous plane. Each plane is symmetrically equivalent to the others and is assumed to be homogeneous. The multi-dimensional spatio-temporal world of physics has symmetric properties. The x-axis can be made equivalent to the y-axis by rotating the figure. However, many phenomena of real life, although multi-dimensional, are neither symmetric nor homogeneous. They have additional properties which call for a new concept of multidimensionality.

The following example illustrates the twisted multidimensionality of real life. When focusing on a physical object, a needle for example, we can explore the object by following a rational route of analysis in the dimensions of

Figure 3.1.   Geometrical lines distorting the visual perspective.

its material constituents, its metal alloys, its surface structure, its metallic properties, its atomic structure, its subatomic layout, etc.; or we can deviate at many points (not everywhere, however) of this ever deeper analysis to a new dimension of exploration. For example, when reaching the analysis of surfaces, we can explore interactions with the atmosphere, chemical reactions occurring at this interface, energy released by these interactions, the nature of that energy, etc. We can also explore the form of the needle and its interactions with other objects, its interaction with air, with a water stream, with electrodes, with ions, etc. We can explore the processes that allow the needle to be built, the machines and the technology behind it, the economic forces that make needles valuable, etc. That is, starting from the analysis of the object "needle," we might deviate into several different "dimensions" of analysis. Jumping from one level of analysis to the next implies a change in the conceptual framework that we employ for its analysis. Sometimes, jumping does not imply a change in the scale that is used to measure space, but it might implicate a change in the scale that is used to measure time. Changing of the conceptual framework on which we base our observations allows us to see a new "dimension" of the object's nature. This curious property of nature, as revealed by our scientific exploration of the cosmos, is what I will call, for lack of a better word, the Multidimensional reality, *Raumzweigung, Kakodiastema*, or the "dimensional ramification of the knowledge space."

These new dimensions might not be infinite in number, equivalent in depth or fine-grindedness, or symmetric to other dimensions. Each of these dimensions has particularities which require a special science for its study. Most of the scientific fields were created to study one such dimension. These multidimensional ramifications of the knowledge are analogous to the cosmic structures visualized by physicists. Distribution of matter in the universe is heterogeneous and forms a matrix or web with autosimilar properties at different scales, from the macro to the micro.

The various modern scientific disciplines reflect some of the dimensions of reality, analogous to the way our mind, forged by evolutionary forces, has been built to reflect reality, as described so delightfully by Konrad Lorenz in his book *Die Rückseite des Spiegels* (which means "The back-side of the mirror"). That is, our eyes and visual system have adapted to capture and analyze electromagnetic waves; our ears have evolved to capture and analyze pressure waves; our brain has developed skills to capture and analyze causal relationships; our intuition reflects in a simplified form, statistical probabilities of interactions among organisms, etc.

Another relevant aspect of the concept of multidimensionality or Raumzweigung is related to the emergence of technologies. For example, when the human mind studies the various types of energy, it stumbles across various problems. Depending on the type of questions asked about the object of study, different dimensions will emerge. When following the analysis of each of these problems, new routes of exploration appear and new sciences, *métiers,* or occupations develop. This relationship explains the birth of mechanics, aeronautics, electrodynamics, electric engineering, electronics, information technology, social communication, etc. The birth of new technologies is a consequence of the human requirement for increased capacity of our limited organs and senses to connect more completely with the world in order to progress or "to dig" into each new dimension of knowledge. New scientific instruments, such as measuring devices and detectors, must be developed to open new frontiers of knowledge and eventually new dimensions of reality. The telescope opened the world of the stars to us, the microscope revealed microbes, the control of chemical reactions led to the discovery of molecules which serve as the building blocks of life, and the forces that bind metals allowed us to follow the actions of electrons.

The unpredictability of bifurcation points at which new dimensions emerge can be exemplified by innumerable cases. I will list just a few here:

- The history of optical devices. Telescopes, first developed to obtain a better view of the enemy or merchant ships, were eventually directed to terrestrial satellites, the planets, and the stars, expanding our view of the universe, the cosmos, and the world of modern physics. With just a few modifications, these telescopes were instrumental in opening a completely different dimension. This was accomplished with the construction of the microscope which allowed exploration of microscopic organisms; this instrument allowed us to see cells.
- The development of devices to measure the charge of electrons in the seventeenth, eighteenth, and nineteenth centuries opened a new knowledge dimension that eventually led to the building of motors, electric devices, and our modern computers in the twenty and twenty-first centuries.
- The study of the chemistry of fermentative processes led to the discovery of unicellular life.
- The drive to understand the properties of magnetic materials led to the domination of electromagnetic waves, to the radio and television.
- The discovery of radioactivity which eventually led to the building of atomic bombs.

The history of the big discoveries of science can be read as a list of discoveries of new knowledge dimensions, whereas the history of each of the multitude of human scientific disciplines and technologies may lead us to the story of ever deeper explorations of a single knowledge dimension. Technological development frequently leads to new branches of the multidimensional reality. When it does, it advances fundamental science.

The multidimensional character of the world allows new disciplines and new knowledge to arise from the study of the very same objects. The physicochemical properties of a knitting needle require totally different means, tools, and laws for their exploration than the study of its mechanical properties. But in both cases, humans develop a scientific discipline for its study. It is due to science that we are able to detect the multidimensional character of the world. Science is not sufficient to discover new dimensions in our world. We need a healthy dose of creativity and imagination. These features are also part of science, but are not exclusive to it.

SCIENCE OF COMPLEXITY

A complex system is a system with many different parts that interact in different ways. Phenomena that emerge from interactions in complex systems cannot be predicted by looking at the component parts alone. For example, the properties of a water molecule can not be predicted from the properties of oxygen and hydrogen atoms, the two components of water. The properties of complex systems force modern scientists to implement certain adaptations, such as reducing the scope of the problem, which is called reductionism. The complexity of a system prevents us from having a complete vision of it and forces us to look for simpler solutions. In turn, we analyze a small piece of the world that is in our view. This reductionist approach to research can be very successful. But if we want to understand the working of complex interactions, reduction of the problem is not feasible or might lead to unfruitful explorations. A story may help illustrate this point. A drunkard looks underneath a street light for his lost car keys. A friend, who reminds him that he lost the keys further down the street, receives the answer that it is dark and he can see only underneath the street light, and for that reason it is the only place where he can look for his keys . . . That is, we need to keep some degree of complexity in our object of study if we want to understand complexity.

A property of complex systems is that a single phenomenon may have various causes and changes in the system might be difficult to achieve. This is also known as the dynamics of webs or web resilience. Another dynamic effect of complex webs are the so-called "unexpected consequences," where a given action, performed to achieve a specific outcome, might produce a completely different or opposite outcome in complex systems.

For example, who could have predicted that sheep, once they were made sedentary and stopped migrating, would increased their period and the amount of lactancy (period during which mothers provide milk to their young); and that this domestication of sheep allowed humans to knit the wool and substitute leather and hides with cloths and fabricate carpets? Who would have foreseen that the emergence of communally defended houses would lead to villages, cities, and modern urbanization? Who could have dreamed of the ways that the car, electricity, the computer, or the Internet changed society and its economics before they were invented and accepted?

History, especially human history, is full of processes that led to results which were unimaginable before they were implemented. Who could have imagined that the discovery of sulfa and penicillin would trigger an increase in the world population? Or that the discovery of electricity would lead to electric light? Or that the discovery of powder for artificial fires would change warfare? Or that the discovery of the earth's magnetic field would transform navigation at sea?

It is rare to find a driver who can foresee the next time his automobile breaks down. He might fix the lights of the vehicle before starting a long nocturnal trip, only to suffer a broken engine because of the lack of water in the radiator. He might repair the brakes, but then have a short circuit that stops the car from starting. Or he might take great care to choose special new pneumatics and then suffer a broken windshield from a stray stone. The automobile is a good example of a complex system. An automobile has many elements that are required for smooth operation. It is evident that an automobile cannot be safely operated at night without lights. It is also evident that a functional and well-maintained engine is required. In addition, a car requires well-functioning tires, brakes, steering, a gear box, battery, clean windows, and many other components.

A complex system, such as the automobile, has many components. All of them require the attention of the conductor. As this is the case of a relatively simple and well-known mechanical system such as the automobile, we might expect much greater complexity when studying and handling economies, galaxies, and social aggregates of living organisms. We know of no unique and infallible prescription to guarantee harmonic operation of complex systems. Continuous attention to multiple details is the only hope to maintain certain control.

Complex phenomena have prompted humans to seek simple models. Simplicity, of course, is desirable. Models are meant to simplify reality to allow for a better understanding. Often, simplification has its limits if we want to maintain relevant properties of the complex system. Finding the right equilibrium between the need to simplify and the need to capture the most relevant features is a challenge of the modern science of complex systems.

THE INTERDISCIPLINARY PERSPECTIVE

When analyzing complex processes, such as social or economic phenomena, certain phenomena are often the "emergent" product of interactions of much simpler processes. For example, macro-economic phenomena are the outcome of a large number of micro-economic processes which produce "emergent" phenomena at the macro-economic level. These phenomena are difficult or impossible to derive at the micro-economic level. However, a deep understanding of macro-economic processes is not possible without good knowledge of the processes at the micro-economic level.

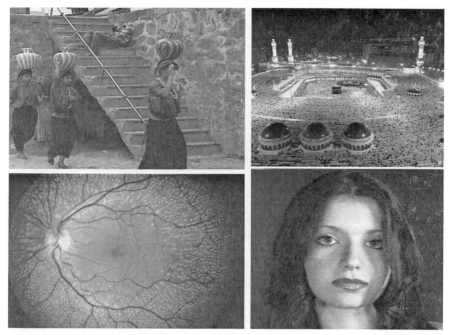

Figure 3.2.   Street protest in Caracas; worship in Mecca; human retina: human face.

The next series of photos (Figure 3.2), all of them representing the same animal, *Homo sapiens*, provide a visual illustration of intra-level dynamics. Each photo represents the same species at different levels of perception and shows different types of multidimensional complexity. These depict the "emergent" product of different types of interactions.

The images show how each scale of observation opens a new world of relationships, laws, and phenomena, even though they all depict the same animal. When studying an organism or species such as humans, we might casually focus on the head and eventually on the eyes of the organism. Examining the interior of the eye with a magnifying lens will reveal a tissue, irrigated by arteries, veins, and nerve fibers. Magnifying one of the nerve fibers, for example, will reveal the existence of neurons, electric pulses, and the frequency-modulated communication system. Focusing on a tiny part of the membrane of the neuron will reveal molecules that interact to produce curious properties at the molecular level.

When traveling from one scale or observational level to the next aggregate level, particular phenomena emerge. When traveling from the cell to the tissue, the retina becomes conspicuous, in the midst of the travel from the tissue to the whole organisms, the organs emerge. When aggregating organs, a self-contained organism becomes apparent. Aggregating a few organisms allows us to detect biologically relevant behavior, such as pair formation and reproduction. Increasing the size of the aggregate of organisms allows for the emergence of new forms of social organization and the existence of mass behavior. The process of jumping levels might be extended at both extremes, at the micro and the macro levels, but this analysis is beyond the aim of the present work. Each of these scales or levels of analysis requires different tools in order to understand them and each is governed by different laws of nature. These laws do not necessarily contradict each other. If they do, they signal failures in our understanding of nature and are likely to be improved in the future so as to achieve consilience.

One aim of science is to understand the mechanisms by which new information emerges. This endeavor has been called the study of emergent phenomena, the study of auto-organization of self-organization, or, simply, the study of the dynamics of complex systems. To really understand emergent properties, however, we must adopt an interdisciplinary approach.

Traditionally, different scientific disciplines focus on different levels of analysis with little insight into the bordering levels. That is the case with micro and macro economics; of cellular, organismal, and ecological biology; of physics and cosmology; and of analytical chemistry and thermodynamics. Yet, at the interface of these disciplines, new phenomena have yet to be discovered. This is an aim of interdisciplinary science.

To start any interdisciplinary endeavor, it is appropriate to focus on differences in concepts and language. These differences often hinder understanding between scientists trained in different disciplines. Only after achieving a common language or means to communicate, can interdisciplinary endeavors be successful.   Table 3.1 gives some examples of different interpretations of the same or similar concepts in three different disciplines whose meanings complement each other.

CONSILIENCE

The Pulitzer Prize winning author and evolutionary biologists Edward O. Wilson ends his book *On Human Nature* with a citation from the Greek tragedy of the Aechylean Prometheus:

Chorus: Did you perhaps go further than you have told us?
Prometheus: I caused mortals to cease foreseeing doom
Chorus: What cure did you provide them with against that sickness?
Prometheus: I placed in them blind hopes

Table 3.1.   Examples of complementary interdisciplinary concepts between three different academic disciplines

Concept	Physics & Thermodynamics	Biology	Economics
Moving force	Gradient	Motivation, drive	Incentives
Scale distortions in a cognoscitive horizon	Waves are seen as chaotic events in small space-time and as regular events in large space-time	Micro and macro evolution differ in emergent phenomena	Micro and macro economics differ in apparent causes and effect relationships
Constant environment	Described by equilibrium thermodynamics	Gives particular evolutionary outcomes	Predictability and the rule of law favor economic growth
Creative cycle	The arrow of time as a product of an expanding universe	Reproductive life histories of birth, reproduction, and death	Creative destruction and technological cycles
Dynamic irreversible progression	Cosmology	Evolution	Technological and cultural evolution
Fractal structures	Sand-strand-geographical borders	Leaf-tree-canopy-forest	Economic transaction-individual motivation-social structures-human values
Freezing	Order emerges at low temperature	Stagnation emerges for lack of environmental change	Rigid structures stop economic growth
Group acting on individual	Fundamental forces of nature influence the dynamics of particles	Social environment guides genetic evolution	Social aggregate forces act through markets on individual behavior
Melting	Order is destroyed by increased	Hybridization and symbioses may produce new life forms	Mixing of individuals and economies may produce innovation

(continued)

Table 3.1. *(continued)*

Concept	Physics & Thermodynamics	Biology	Economics
Order	temperature	Adaptive structures	Specialized social economic structures
	Negative entropy		
Parasite		Bloodsucker	Free-rider
Progress	Side-reaction	Adaptations with increased energy efficiency	Increased productivity and economic efficiency
	Negentropy increases energy efficiency		
Annealing	Re-programming neural networks	Assortative sex and hybridization	Business contracts and mergers
Revolution	Phase transition	Speciation	Change in business paradigm
Stochastic blocking or jamming	Excess noise freezes the system	Excess variability stops evolution	Excess individual deviations blocks progress (traffic jams)
Stochastic resonance	Noise improves signal detection	Noise improves adaptive efficiency	Risks improves profit
System—hypersystem	Stars—galaxies—universe	Species—population—metapopulation	Business—markets—economy
Threshold		Edge of chaos	Barrier to entry
Threshold soup	Activation energy	Evolution driven by sexual selection	Emergence of innovative societies
Umwel	Turbulent flows	Ecological environment	Economical environment
	Border conditions		

This ancient statement, which seems to be an astonishing truth to our contemporary ears, ends an exploration that originates from the biological nature of *Homo sapiens'* most salient features, such as ethical behavior and religion. Thus, I start where E. O. Wilson ended, meaning that both his view of Human Nature, and his approach to the consilience of the sciences (Edward O. Wilson, *Consilience*. Vintage Books, 1999), should be used as a basis upon which the construction of any new science, especially the modern human sciences, can be advanced. But, this advancement has to take care to maintain consilience with the physical, chemical, and biological sciences. By consilience, we refer to the harmonic overlap of the various scientific disciplines, starting from physics and chemistry, bridging biology and anthropology, reaching into psychology and sociology, and affecting the foundations of philosophy, ethics, and religion.

For example, only one of the drawings of networks in Figure 3.3 is consilient with neurophysiological phenomena and psychological sciences. Although we do not know all the treads leading from the morphology of neural networks to consciousness and other psychological phenomena, we know that we achieve consilience if our knowledge of neuron networks does not contradict our understanding of neurophysiology and psychology, and vice-versa.

Consilience can be regarded as a falsifiable condition of interdisciplinary science applied in a complex context. It seems to be an especially useful concept if we want to fuse the social sciences with natural sciences, using different tools to explore different phenomena, but bridging disciplines without losing sight of each individual discipline. Consilience among the sciences means

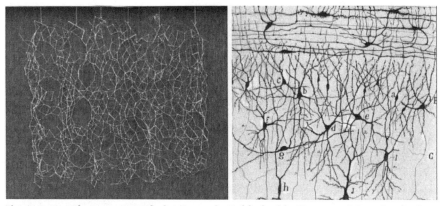

Figure 3.3. Abstract networks by Gertrude Goldschmidt (Gego); and Drawing by Santiago Ramon y Cajal of the first and second layer of the olfactory cortex from the hippocampus of a child.

that any theory built to explain phenomena observed at one level of reality, tackled by a group of scientific disciplines, has to be consistent with, or at least shall not contradict, disciplines at other levels of reality.

For example, a physical theory of the cosmos is unscientific if it uses assumptions that contradict scientific theories of particle physics. Similarly, any theory of chemical processes has to reconcile its assumptions with what is known about the quantum mechanics of the forces acting on atoms. No biological theory can be considered scientific if it assumes realities that contradict our chemical and physical description of the phenomena. Similarly, no human or social science can be considered a "science" if it is not consilient with biological, chemical, and physical views of the world.

This does mean that science at higher levels of complexity, or at different knowledge dimensions, can not analyzes phenomena that are not present at lower levels of complexity or at other dimensions of reality. Each scientific discipline requires new tools to study and understand properties of the system under study. But many features at a higher multidimensional level are just emergent properties of the interacting features of a lower level. The consilience between disciplines guarantees that these emergent phenomena will eventually be recognized as such. Consilience, in true Popperian fashion, means that theories at a higher level of complexity should not be inconsistent with those of lower levels of complexity, unless they aim to falsify these theories. That is, consilience might be proposed as a new way of falsifying complex hypotheses and the degree of consilience between disciplines might serve as a measure of the adequacy and scientific value of these disciplines.

Consilience draws on the belief that all phenomena on earth are interconnected. That is, mathematics can explain physics, physics underlies chemistry, chemistry and physics form a basis for the understanding of biology, and sociology and psychology are an appendage of biology, the science of the study of the living. As humans are animals, anthropology is a branch of biology and when applying these disciplines to social behaviors, sociobiology emerges. This logic also applies to specific disciplines, such as sociology, which is dedicated to the study of social interactions, or economy, which is focused on specific types of interactions among individuals of a single species. At present, a huge gap separates natural sciences (mathematics, physics, chemistry, and biology) and social sciences (anthropology, archeology, sociology, psychology, and economics). Consilience is a tool that may help bridge this gap. This is especially true when trying to build models or select between models that aim to explain a family of phenomena.

SIMULATIONS AND MODELS

Models are an efficient way for the human mind to produce simplified manageable versions of complex reality. This allows an individual to foresee the consequences of its actions and to plan accordingly. Models are especially useful when taking complex problems. Parts of reality are so complex that the human mind cannot grasp its entirety. That has made simplifications of reality the only route available to handle complex systems. Many properties of a complex system are lost when we simplify it too much. Happily, with the emergence of inexpensive, accessible, and powerful computing, ever more complex modeling has increasingly been performed with the aid of computers. These computer aided models are called simulations.

Simplified models are useful since absolute truth is not achievable. We might only grasp parts of reality when we simplify and reduce our scope of attention with a model of reality. Our brain is so much smaller than the real world that only crude and synthetic abstractions of reality, such as those achieved with models, are practical. Absolute truth as a perfect reflection of reality is unachievable. Relative truth in the form of synthetic abstractions (models) approximates specific realities which are more achievable than absolute truth.

The history of science, and even the history of life, can be viewed as a history of devices for model building that allow a better understanding of increasing levels of complexity or higher levels of organization. Although the exact evolutionary sequence of events that led to the creation of the minds of modern humans remains unknown, a pseudo-sequential, punctuated history of the development of mental and analytical models during the evolution of life should include the following aspects:

- The first living beings developed an abstract model or code of their body coded in a genome made of DNA that provides instructions for constructing their organism. The emergence of the genome and its genetic code reflects the need to survive in a given environment by using a device for storing information.
- Organisms evolved a central nervous system. This organ allowed for modeling of movement by the organism in three-dimensional space.
- Once a nervous system had evolved, it allowed organisms to build representations of themselves and the environment.
- Self representation allowed the emergence of consciousness. Most mental processing of spatial perception and dynamic analysis is performed unconsciously, using inborn neuro-physiological mechanisms. We become

conscious of our unconscious processing of spatial perception when we detect its limitations, as beautifully exploited by several artists, such as Martin Escher. Modern psychology explores these limitations to gain insight into the workings of our brain, producing interesting new designs. One example is the Ouchi Illusion (Figure 3.4) where static lines give the impression of movement.

- More complex analytical mental processes allowed for the estimation of proportions, quantities, risk, and probabilities. This allowed the human mind to develop geometry, algebra, and calculus.
- The adaptive pressures suffered by our thinking tool, the brain, include the fact that it is immersed in a social context, which forced our mind to evolve the capabilities to model the social context. Our mind learned or evolved

Figure 3.4.   Ouchi illusion by Kitaoka A and Murakami.

the ability to look at the social world, first subjectively (ourselves), then with respect to family (kin relationships), and then through the minds of others fomenting the emergence of empathy. Empathy builds on the capability of modeling the mental models of others.

- Interactions of mind and body during primate and human evolution molded the human mind, as it produced simple and efficient decision rules for analyzing complex bio-social contexts. One such set of rules is based on a psychological model revealed by Freud's "Psycho Analysis" which expands basic sexual drives to form a more comprehensive libido, producing a potent psycho-social motor that guides our motivations and actions. Another simpler psycho-social model is based on the family structure and is narrated by "Transactional Analysis" through the Child-Adult-Parent roles that the human brain is biologically prepared to play.

- A series of evolutionary paths lead to the recognition of a subject as separate from the environment, the existence of others which allowed for the emergence of empathy, and the limits of our minds, which led to scientific models that continuously search for anchors in reality.

- Building on abstract model of family and sex, our brain produced highly sophisticated models for abstract visualization of the interplay between ourselves, others and the environment in the past, present, and future. Some of these models were branded as religious or transcendental ethical constructs. These highly abstract ethical models produced concepts of "the good and the bad."

- A global and highly technical world, aided by modeling tools such as the computer and internet, produced world markets in ideas, goods, and capital, which allowed global models to emerge. Some of these global models, focused on very specific aspects that affect humanity, are highly formalized, such as those used to plan the destruction of the world by atomic bombs, or those analyzing macro-climatic phenomena, such as global warming.

- The development of digital information systems, which work much faster than human brains, allowed for the emergence of individual-based numerical models, agent-based simulations, genetic algorithms and other gadgets of artificial intelligence that are able to produce models that often outperform our mental models in speed, accuracy, breadth, and relevance.

Some aspects of the evolution of science are analogous to the evolution of the human mind, possibly with some homologous elements. The history of science can be viewed, in the context of the different dimensions of knowledge exploring different aspects of reality, as a history of evolution of ever more sophisticated and accurate models, which opens up new dimensions of reality with new technology and new conceptual models. These models are subject to adaptation and selection, which continuously improve their ability and accuracy to explain and predict new and old natural phenomena. Such a history, centered on the evolution of scientific models, can be depicted as follows:

- The most basic scientific models used by primitive humans improved individual survival, driven by basic natural selection. For example, mental models improved hunting by allowing for testing, prior to application, of an alternative plan of attack. This is exemplified by neurophysiological studies which showed that cats dream of hunting mice, suggesting that their mind constructs a scenario where the cat is hunting. Cats, dogs, and many other animals build cognitive models. Thus, primitive scientific models dealt with space, geometry, and cause-effect chains.
- Mental models allow for the detection of imperfections in our perceptions. We discover that reality, as we see it, may be a game of shadows as pictured verbally by Plato's man in the cave, looking at the shadows of the outside world, or pictured visually by the drawings of Rob Gonsalves. To overcome this limitation we rely on instruments and data from a variety of sources to validate our models.
- The need for validation requires better formulated scientific theories that allow for falsification and numerical models that allow for quantitative predictions that can be easily disproved.
- Further development of scientific models produces meta-science and the search for consilience. Models transcend individual disciplines, but look for webs of anchors to contact reality.

Not all human cultural developments lead to science. An alternative development that leads away from science includes some of the following:

- Our mental models become so efficient at explaining our experiences that we stop making contact with reality. The virtual world becomes reality. This allows the mind to construct models that are not bound or constricted by reality. The thought process becomes surreal and fantastic. The freedom gained from reality allows for explorations of our fantasy that lead us to mental worlds that seem infinite and astonishing. This route is taken, among others, by parts of literature and poetry.
- The construct of models and rational mental frameworks that do not require continuous validation with reality allow for very fast developments, producing an impressive amount of intellectual work. It is sufficient for scholars to root their arguments and logical chain of thought with some widely respected authority (preferably on observations of Aristotle or Plato, as in many humanis-

tic disciplines). This habit allows for the existence of scholars that are not interested in contacting reality since the founding authority of the discipline did, despite the fact that reality may now be perceived by humanity in a radically different way, thanks to the fact that our means to explore and query nature are incomparably more sophisticated and accurate than what Plato envisioned.

- The model may become a substitute for real life. Our mind creates reality and dogma, giving birth to religion and the post modernism dissolution of the limits between illusion and reality. Modeling without contacting reality with rigorous methods is the basis for non-scientific thought. The lack of experimental validation of mental constructs, which allowed for the appearance of a large body of literature, blurs the distinction between reality and fantasy, dreams and real history, experience and imagination.

To summarize this method of analyzing history, I will highlight a few important steps that led to the evolutionary emergence of enlightened scientific thought. Accepting that human progress is discontinuous and that no linear sequence is likely to explain the adaptive challenges that the human mind encountered during its evolution, a rough approximation of some important steps in the evolution of modern science would include:

- *Survival*: Development of randomly acquired features for survival expressed by a randomly drifting amoeba
- *Search*: Foraging for food or shelter
- *Planning*: Creating images or copies of reality in order to plan foraging strategies that optimize actions, such as when choosing the shortest path
- *Thought*: Creating actions in an imagined space
- *Conscience*: Including the thinker as an actor in the imagined reality
- *Realism*: Acknowledging that, although the habit of substituting reality by imagined reality (idealism) is important, it produces approximations to reality and not a substitute of it
- *Science*: Interplay between deductive analysis, experimentation, reductionism, and induction, leads to a dialectic relationship between reality and imagined reality
- *Scientific Theory*: An abstraction of reality that allows for efficient planning

THE EMERGENCE OF META-SCIENCE

Different disciplines of science are focused on different levels of development. Employing the terminology used to describe the evolution of insect societies, we postulate various levels of scientific endeavor.

- The first level would constitute the prescientific level, where human endeavor forces the mind to be rational, i.e. to follow logical steps of thought, such as was done by Pythagoras, Socrates, Aristotle, and some of the ancient and present day philosophers.
- A second step, defined as the parasciences, is when human rational endeavor is dedicated to describing nature. In this context, parascience was performed by all natural philosophers, natural history researchers, and present day taxonomist and anthropologists, among many others.
- A third level involves the true science or euscience, as popularized by Galileo, in which both theory and experimentation are important parts of the rationalization of reality, but where experimental results hold priority over theoretical concepts. Disciplines which fulfill the criteria of euscience are most of physics and chemistry, molecular biology, and population biology, among others.
- Other human rational heuristic endeavors have either no experimental confirmation of theory or no falsifiable theory, yet they call themselves scientific. Such endeavors can be classified as pseudoscience.
- Recently, computer-aided theoretical models provide falsifiable predictions in meteorology, turbulent phenomena, cosmology, and biological evolution. Such an endeavor is not euscientific as the theories underlying the numerical models are complex constructs and can not be falsified as a whole, but each of its parts can be falsified. Falsification and theoretical buildup is accomplished using computers and artificial intelligence. Such activity can be classified as metascience.

The most developed metasciences today are probably those involving meteorology, hydrology, and fluid dynamics. Another example includes cosmology, social simulations, artificial intelligence, complexity theory, and sociodynamics, which where employed in an attempt to create a conceptual framework for the interdisciplinary study of subjects spanning biology, sociology, politics, history, economics, and other disciplines. The framework of metascience is based on principles that are eventually reducible to or consilient with the laws of natural science. The robustness of this framework might be tested with simulation models that allow performing virtual experiments with complex systems. Based on these results, specific predictions may be formulated which in turn may be subjected to further tests. In this way, mechanical physics leads to the development of meteorology and global climate science and quantum mechanics leads to the development of theoretical chemistry. Similar attempts to build a metascience of biology are more recent (e.g. artificial life) and are not easily falsifiable with natural experiments. Another example is the application of game theory or study of the effects of specific decision making mechanisms on the dynamics of assemblages of individuals, to an ever increasing range of phenomena, as common now in sociobiology, sociology, and economics. This metasciences provide, among other things, explanations for the emergence and maintenance of societies. A methodological framework for metascience is still lacking, but it may help bridge the gap between the natural and social sciences.

Metascience relies heavily upon the contemporary activity of meta-analysis and various forms of data mining. New analytical tools that employ computers allow for analysis that is completely inaccessible to our unaided minds. Thus, the development of ever more powerful tools for analyzing complex assembles of data will allow for ever deeper expansions of new scientific disciplines. One such discipline is related to the study of the dynamics that underlie social dynamic phenomena which might be called sociodynamics.

STUDYING SOCIAL DYNAMICS

Many complex dynamic phenomena, which were long thought to be beyond the realm of hard science, have entered the realm of metascience. This allows establishing a sound scientific basis for our understanding of the interacting web of biological, social, and economic behaviors. The study of social irreversible dynamics is fast approaching the characteristics of a quantitative experimental interdisciplinary science and draws heavily on evolutionary biology, thermodynamics, physics, and computer science. Many such efforts aim to bridge the gap between the social and natural sciences by maintaining consilience among the disciplines and utilizing interdisciplinary approaches. Interdisciplinary and multidisciplinary endeavors are full of pitfalls that can entrap scientists that step beyond the familiar terrain of their specialized discipline. However, it is only with interdisciplinary and multidisciplinary research that complex problems, which are relevant to the future of humankind, will be tackled. Thus, we must advance through a jungle of complex knowledge, even though we are well aware that we will encounter pitfalls. There are no alternatives to advance our understanding of humankind. We should be aware of the heuristic value of a theory: It is more important that the theory allows others to advance knowledge, even if this requires disproving a theory, rather than being correct in all its details (although this might be important for other reasons).

A common misconception of natural scientists is that social sciences do not use sophisticated mathematical tools. On the other hand, social scientists and many lay persons often have the impression that most phenomena in natural science are described by deterministic linear laws. Natural science is full of stochastic phenomena and phenomena which are non-linear in nature, i.e. the relationships between underlying variables cannot be described by simple formulas. This erroneous belief in the absolute nature of all laws of natural science has brought a dangerous simplistic interpretation or translation into the domain of the social sciences, which discredits the efforts to translate methods from the natural sciences to social sciences. For example, a simplistic interpretation of Darwinian forces ruling human social behavior equates economic market behavior with the "survival of the fittest" rule. This ignores the fact that biological evolution is highly stochastic in nature and is much more concerned with the survival of the luckiest than of the fittest. The main difference between natural science and others is that natural science deals with relatively simple systems and thus can apply the scientific method more rigorously. The unconditional application of the rule "experimental evidence overrules any conceptual construct" to the study of social phenomena produces a euscience of human social behavior.

FROM INDIVIDUAL CONSCIOUSNESS
TO SOCIAL KNOWLEDGE

All living creatures have a basic level of consciousness. They receive information from the outside world and differentiate that information from that which they receive from within they organism. It is the quality of information processing which defines various levels of consciousness. A level that is recognized as consciousness by most people might include a sensory system in the organism that can differentiate between external and internal stimuli and can perceive itself as being distinct from the surrounding environment. A next step on the ladder of consciousness is the provision of mental tools that enable the organism to see itself embedded in its environment and to model, imagine, or predict the outcome of specific actions in its environment and the effects of the environment on itself. The next step, which leads us to the definition of consciousness given by John Locke in 1690, would include "the perception of what passes in a man's own mind." A further step, which leads us to the definition of conscience as used by the Encyclopedia Britannica, is "a personal sense of the moral goodness or blameworthiness of one's own conduct, intention, or character with regard to a feeling of obligation to do right or be good." The use of consciousness allows social structures to reap benefits for all (or most) of the participants through the building of synergistic relationships.

Societies promote interactions between individuals which are often competitive, especially males. The German chemist Justus von Liebig summarizes the value of competition in science as follows, "I seldom have a good idea, but if someone else comes up with one, I immediately have a better one." This shows how important completion is for science. On the other hand, modern society is increasingly dependent on the cooperation of its citizens, as is science. The average number of authors on scientific papers is increasing year after year. The lonely scientist that advances human knowledge while isolated in his lab and communicating only from time to time with his colleagues, is an image that conforms less to reality as the years progress. The complex network of collaboration in megaprojects, the international and interdisciplinary nature of these collaborations, and the impact of these megaprojects on science and on society, make them more common and more important. Modern science is not only an activity performed by isolated scientists, but it is increasingly an activity that depends on of the work of others.

A large part of the success of science and its armies of scientists that advance the frontiers of knowledge is dependent on information. No advances in modern science are possible without an efficient flux of information between generations (through books and papers) and between scientists that produce new information (through correspondence, publications, meetings and congresses). An abundance of new knowledge is achieved through the internet that allows building large databases and the increasing number of specialized scientific journals freely available on-line. These features are part of *The Information Society*. Here the tool for creativity is not the mind of an individual scientist, but the collective brain power of a community of scientists.

Collective or aggregate knowledge is of fundamental importance when considering the advancement of metasciences and the efficient search for consilience among the emerging sciences. Future science will be constrained in great part by the degree and type of communication developed by the scientific community. These constraints can only be understood by first understanding the interactions between the individual and society and by recognizing the properties that emerge from these dynamics.

Globalization of knowledge should lead to the globalization of conscience defined as a collective sense of the moral goodness of the conduct of individuals and societies with regard to the common good of humanity.

This process of fusing individual consciousness with social consciousness has begun, but is far from complete. Modern society and the scientific community are developing new institutions to solidify this process. It seems that humanity has a way to go before settling into a stable, equilibrated, and sustainable society. The scientific community and the institution that it is creating, serve as a kind of forefront for social experimentation. The internet was developed for scientists to efficiently communicate and it is now part of every day life for most citizens in developed countries. Other means of communicating, harnessing, and managing information those were originally developed by and for scientists are also being used in increasing measure by common citizens. Peer review, open access to information, decision making by consensus of experts, search for broad and diverse interdisciplinary expert advice, and reliance on experimental proof rather than on the beliefs of a majority, are features of the scientific community that may spill over to other parts of society.

FUTURE QUESTIONS FOR METASCIENCE

New disciplines are likely to emerge from interdisciplinary metascience. The list of these new scientific disciplines is probably very large. If we take the *Raumzweigung* of reality into account, attempts to list possible new questions for science are futile. Here, I will present a few research questions that might emerge if humankind manages to bridge the gap between social and natural science, or between natural and moral philosophy. This list aims to portray areas of future scientific activity to stimulate the reader's interest.

- What are the mechanisms that rule the dynamic interactions between economics and politics?
- How do different motivation thresholds (salary gradients for example) affect economic growth?
- What are the differences between engineering and science? Do those different attitudes affect the political and social world?
- What is the impact of TV and other media on the social dynamics? News reporters distort reality as they present the exceptional as the common and the common is ignored. Events are presented without reference to the likelihood of its happening confusing the common with the exceptional. The future of society will be affected by the way information is gathered, stored, and distributed.
- Hope and leadership have been inseparably interwoven. The most extreme example of promises of future happiness is provided by religion, where death or afterlife is used as the time horizon for any return to present sacrifices. New time perspectives of political action might produce new social organizations with decentralized leadership. If so, science will play a central role in the analysis of these new realities.

- Social illness and widespread psychological deformations may constrain economic growth that is affected by ecological settings (tropics vs. temperate, for example).
- Bioeconomic questions. For example, parasitism can be studied as a form of accumulation of wealth in humans and governments can be perceived as modulators of natural selection by favoring gradual change, focusing on long-term time horizons, and stopping or reversing the selection of specific traits.
- How do rational individuals should perform in an ocean of irrational individuals? What is the optimal algorithm of action for rational operators, acting through irrational operators with rational bits such as politicians, but constraint by actors guided by affective forces such as voters?
- How does democracy and information interact? A uniform distribution of information optimizes democracy as it maximizes the efficiency of decision-making. A heterogeneous distribution of information produces a deficient democracy. Attempted solutions to this problem are: 1. Distributing information broadly enabling the understanding of the relevant issues; 2. Selecting capable voters by restricting democracy, according to issues, to those that are informed; 3. Developing an aristocracy. Other solutions will have to be developed.
- How can egoism be handled more efficiently for the benefit of all? More examples of known methods need to be developed. The classical systems that perform this trick are: The Wali system practiced during the Ottoman Empire which legalized corruption by distributing governmental posts by means of auctions. The free-market capitalism where intense competition melts individual and social interests.
- What is the effect of different information transmission systems on social structures? For example, what kind of family tradition will prevail in a given culture? Two types of information transmission in cultures spring to mind: Chinese entrepreneurs are based on vertical transmission of information and family networks, whereas the global info-society, such as the global science publishing system, is based on meritocracy, assigning the best prepared person to a specific task. How does information flows in each system and at what speed? What are the constraints of family networks in business?
- How should the efficiency of different types of synergy be measured?
- How should cultural features be quantified? A fine quantitative assessment of cultural traits would eventually allow us to; for example, assess the history of the proportion of society dedicated to knowledge acquisition and the economic cost invested by society on this endeavor. For a broad understanding

this history has to include non-human societies, such as ants, and human societies; expand to primitive human cultures such as Yanomamo and ancient Egypt, and then tackle modern developing and industrialized countries.

- Future social science will need to provide quantitative and predictive models to explain, in detail, fundamental social phenomena such as the poverty trap, diffusion of innovation, the relationship between education and economic growth, and the emergence of macroeconomic rules.

Chapter Four

What Science is Not

Many rational qualities are often associated with science. Some of them may occasionally add some value to the scientific enterprise in a given area or at a given time, but none of them are essential parts of the scientific method. Some of these include:

- Scholastic logical thinking and rational deduction without experimentation are important aspects of theory building and are essential to science; however, rational logical thinking without regard to experimentation and use of individual consciousness as a final arbiter are non-scientific endeavors.
- Expressing or believing in certainty without doubt is unscientific. Although some theories might have achieved high levels of certainty, doubt or a skeptical attitude is part of science. Any theory may be disproved with new facts.
- Simple theories draw on chains of casual relationships or causality to explain the phenomena of nature. More complex theories are based on stochastic statistical theory and are more prone to use casualty as an explanation. Truth seems to have a stochastic component.
- Symmetric and clear mathematical formulations are appreciated in a good theory; however, mathematic expression of ideas is not sufficient to demonstrate veracity. Experimental verification is essential, although mathematics might aid in this endeavor. Mathematical theorizing without regard to experimental verification is non-scientific. A classical case of this pitfall is the study of the Kabala.
- Rigorous data handling is often thought to be sufficient for qualifying the endeavor as scientific. Yet it is refutability that differentiates the theories derived from this data as scientific or not.

- Separating practical errors from theoretical errors. Often, a theory can not predict practical outcomes of the phenomena it aims to describe and the lack of falsifiability is assigned to practical errors that have nothing to do with the theory. This has happened in some cases but ignoring repeated refutations by experiment is non-scientific.

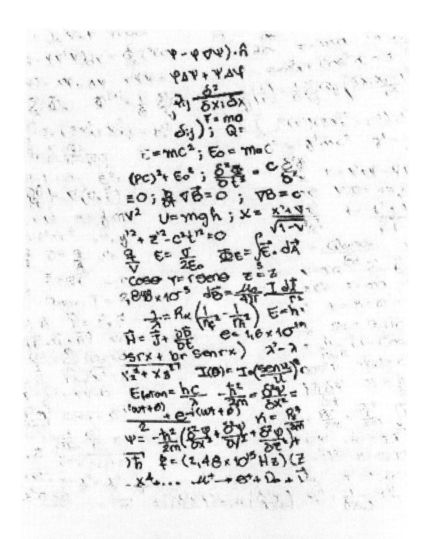

Figure 4.1.   Drawing by Lorena Jaffe.

- Words pronounced by Albert Einstein are probably more valuable than those pronounced by Peter Schmidt. But, even Albert Einstein was wrong many times. Thus, expressions from authorities cited in academic journals, however impressive, are not a substitute for rigorous proof of facts. Papers that base their conclusions only on authoritative citations are non-scientific.

The most dangerous impostors are probably pseudoscience and verbal artists or charlatans that claim to do science. In the popular consciousness, these are often confounded with true science. We will explore these disguises further.

WHAT CHARACTERIZES PSEUDOSCIENCE?

Some people want to collect the fruits of scientific success without actually performing scientific research. This can be accomplished by capturing some of the fame and notoriety surrounding the scientific method. This is an anti-scientific or non-scientific endeavors disguised as science. We call these endeavors pseudoscience. Based on the articles *"Pseudociencia y cultura de masas"* by J. Medin & E. Núñez, of the University of Puerto Rico, I present a reduced but representative list of pseudoscience characteristics:

- A lack of falsifiability or the construction of an irrefutable hypothesis. This is the case for most parapsychologists who claim that extra-sensorial faculties can only be observed if people believe in them. Thus, experiments can not be designed to test the veracity of parapsychological claims.
- Search for truth by examining sacred texts. If a text, such as the Bible or astrological tables, is assumed to be true without further recourse to falsifiability, all conclusions drawn from the analysis of these texts are also not falsifiable. This characteristic is common among irrational followers of great thinkers and scientists. Some psychoanalytical schools, homeopaths, astrologers, Marxists, Keynesians, Darwinians, Aristotelians, and others draw their wisdom from the writings of the founder of their discipline. The founder might have been following scientific methods, but the irrational and uncritical acceptance of these texts makes the followers non-scientific.
- The interpretation of scenarios: Crystals are used to cure illnesses, but the success of this cure is based on the supposition that unknown physical forces affect our physiology and that correct manipulation of the crystals will guarantee a correct outcome. Any failure of the method is then assigned to incorrect proceedings.

- Mixing metaphor and reality: Astrologists, card readers, and other nigromants base their search for truth on finding similarities between signals and reality. Any mismatch is blamed on an incorrect interpretation and will not invalidate the method.

- The majority dogma: Even if a great number of people believe that the earth is flat, this does not falsify the fact that the earth is round. Scientific facts have very little to do with the logic of the masses. Magicians, although clearly not charlatans since they recognize that they are performing tricks and proclaim to be selling illusions, aptly use this argument. They insist many times that the pocket is empty before extracting a white duck from it. Pronouncing the word "empty" many times does not make the pocket empty. Although for human intuition, lies told many times become more believable.

- A lack of a reliable conceptual framework: Homeopaths base their practice on the thought that similar causes cure similar ills, That is, if you have fever because of Malaria, for example, a potion that produces fever should be appropriate to control your illness. These pseudo-rational conceptual frameworks are not based on any testable theories and thus can never be considered to be scientific.

- Anarchic methods: Changing the subject of our attention or the method of study frequently makes the subject of our study unfalsifiable. Many charlatans resemble a lawyer more than a scientist when they explain their pseudo-science. The abundant use of poorly defined concepts makes it impossible to follow their argument in a logical manner, which makes it hard to refute them.

- What looks like a mystery must be a mystery: A romantic idealization of information generally leads to a false conclusion. The mysteries of Loch Ness and the Bermuda Triangle exist because we do not understand them. This nebulous description unlocks our fantasy and makes the subject interesting. Any search for the truth is unwelcome.

- Negligent handling of evidence: Constructing histories based on randomly picked evidences leads to irrational beliefs. Also, confusing correlations with causal connections or assuming that if you see something three times it must be statistically relevant, are typical procedures of pseudo-scientists.

- Immunity to criticism: Fanatics are immune to evidence that conflicts with their beliefs and most pseudo-scientists are immune to criticism. They do not want to know the truth. They are happy with what they believe and have no interest in furthering their knowledge.

- A lack of progress: Any heuristic activity that has not advanced our understanding of nature or ourselves after establishing itself is clearly not science. The practice of homeopathy and parapsychology has hardly changed over the last two centuries, whereas chemistry, biology, and astronomy from a few years ago are completely different from the current landscape of the respective fields. It is the measurable advance of knowledge that makes science such a powerful heuristic tool. Any human activity that does not show such progress can hardly be called a science.

- Irrationality: A lack of logical thinking, difficulties in establishing chains of causation, and neglect in constructing coherent abstract models often contribute to pseudoscience (See also *The Enemy Within, Stuart Sutherland, Penguin, 1994*).
- Confusing method with science: Following a specific methodological framework and disqualifying any approach to reality that does not follow that framework as unscientific, is not science.

THE ART OF CHARLATANRY

Exceptional communicational abilities are important to convince fellow humans of your ideas. Verbal arguments are the natural method used to convince a child to obey social rules and amend their behavior. It is a deeply rooted human instinct that affects the way we think and make decisions. Of course, some people are more gifted than others in using verbal arguments. The use of verbal arguments without regard to empirical evidence is called charlatanry. Charlatans were successful in attracting large number of followers in ancient times and continue to be so in modern society.

The incredible success of science has its drawbacks. The taming of atomic forces led to the building of the atomic bomb; and the discovery of penicillin and other antibiotics led to cures for many illnesses that allowed people to live healthier lives by avoiding infections. These and many other achievements of science engender the feeling that science is all powerful. This causes many to draw on this prestige and power without engaging in the penuries of practicing science. Charlatanry is not new and will probably never be extinguished; but, today, many charlatans sell themselves as scientist and pseudo-scientists. It seems appropriate to build a classification of charlatans that might help us debunk them as pseudo or non-scientists. Charlatanry includes practitioners of the following activities:

Parapsychology: The pseudo-scientific pursuit of paranormal phenomena, such as the search for telepathy, clairvoyance, precognition, telekinesis, and other manifestation of putative extra-sensorial capacities.

Astrology: Relates to the study of the supposed effects of stars and planets, according to their position in the heavens, on human life and other terrestrial events.

Ufology: Studies the apparent visits of extraterrestrial life to our planet.
Creationism: Posits that all animal species were created by divine decree and
    not through biological evolution.
Pseudo-economy: Uses the dictates of a given ideology to explain economic
    phenomena. Fanatical Marxists and uncritical supporters of an unregulated
    free market belong into this category.

History has taught us that some pseudo-sciences, such as alchemy and hyp-
nosis, transformed themselves into science (chemistry and psychology, respectively). Yet, these events occurred through the recognition of failures of
the theory to explain empirical, fomenting the adoption of the scientific
method.

THE LIMITS OF REASON

*"Human reason . . . is called upon to consider questions, which it cannot
decline, as they are presented by its own nature, but which it cannot answer, because they transcend every faculty of mind"*

—Immanuel Kant

"Truth is more likely to arise from error than from confusion"

—Francis Bacon

*"The man who is certain he is right is almost sure to be wrong; and he has
the additional misfortune of inevitably remaining so. All theories are fixed
upon uncertain data, and all of them want alteration and support"*

—Michael Faraday

It is important to note that science and rational thought also have limits.
The limits of the rational mind are plentiful and it is essential to be aware of
its limits for the sake of rational thought.

As has already been described in the text, a fundamental assumption in
science is that our minds are bounded and that there are limits to rational
thought. But science without a mind guiding it can not exist. Thus, a deeper
understanding of the workings of our rational minds and of our mind in general is indispensable if we want to understand science as a phenomenon in
itself.

Science can not be separated from humans. After all, it is a human invention. Thus, understanding and exploring the limitations of our mind regarding

rational thought might help in avoiding some pitfalls. Some of these pitfalls are (adapted from *Pseudociencia y Cultura de Masas.* by J. Medin y E. Núñez):

- A tendency to underestimate the probability of coincidence: A person can think about a tragedy and, a few days later, the tragedy happens. We dream of meeting a long lost acquaintance and then we find the acquaintance on the street. How is this possible? It can not be a mere coincidence. Yet, it was a coincidence. Even trained scientists underestimate probabilities associated with coincidences. The classical example is the probability of finding two people with the same birthday in a party of 30. Most people assume that this probability is very low. In fact, it is very high (71%). Thus, each time we go to a party, we should not be surprised upon finding coincidences in birthdays, names, or family characteristics.
- A lack of appreciation of randomness: Most people, when asked to write a series of random numbers, will avoid repeating the same number consecutively and will try to choose numbers uniformly from the range allowed. Yet, random numbers are most often than not non-uniformly distributed and a series of consecutive equal numbers is very likely to occur.
- A tendency to jump to conclusions: The lack of an accurate sense for statistical probabilities makes our mind jump to conclusions after a few events have been analyzed. The need to build comprehensive models quickly, in order to act rapidly, is a hindrance to scientific thought. For example, the fact that the conditions for life might exist in other planets, often leads us to assume the existence of extraterrestrial life. Yet, many steps of logic remain to be addressed before we can reach this conclusion with scientific rigor.
- A tendency to perceive order in random arrangements: The adaptive value of our mind to build models and draw conclusions is clear. We needed to design an escape route to fool a running tiger or outwit a hyena when hunting for meat. Speed was more important than accuracy. Yet, for science, the priorities are not the same. One such difference can be found when we observe randomly produced structures or series. Our mind is quick to find regularities, even if they do not exist. Even worse, once such a pattern is perceived, it is hard to make the mind change this perception. We will insist on seeing these patterns, even in other circumstances. For instance, we see familiar animal and human forms in the constellations of stars. Rocks, stones, mountains, and clouds show faces or animal figures that we recognize. Certain signals of nature are associated with success or failure in sports and politics. Even some apparently sober scientists saw channels on the surface of Mars before satellite photographs corrected the psycho-visual distortions.

- A tendency to detect spurious correlations: Similarly, correlations between events suffer from a similar lack of statistical rigor. Superstition is the logical outcome of an over sensitive mind with regard to the detection of coincidence. One or two events correlated with a tragic accident are sufficient to establish a life-long correlation between them. Phobias, irrational fears, and superstitions are much more common than most scientists accept. Barry Singer performed related experiments under controlled conditions. A group of people were given problems without solutions. When each of the subjects proposed a solution, the experimenter randomly informed them that they were correct or incorrect. With this false and random information, subjects produced the most fantastic theories. Even rats, trained to press a handle in order to obtain water, often associate certain body position or certain sequences of action with the reward. As a result of this association, they develop superstitious behavior.
- A propensity to ignore unfavorable evidence: If you are a lottery player, it is very likely that you remember the few occasions you won, if ever, but will have difficulty summing up the amount of money that you spent playing over the last few years. These mental shortcomings enable future-tellers and fate-viewers, who barrage their clients with diffuse and verbose information, to be sure that they will mention a startling coincidence and that the client will henceforth only remember the acquaintance. Fanatics are famous for ignoring unfavorable evidence. Most of us have a fanatic vein. Intellectual discipline is required to accept facts as they are and to rearrange our theories and models of reality if they are not supported by the facts. Critical thinking is not a given virtue of the human mind. Considering multiple hypotheses and being prepared to discard the favored hypothesis when facts do not support it is even hard for scientists.
- A constructive and selective memory: Judiciary trials rely very much on the memories of victims and witnesses to place guilt on the criminal. Yet, more controlled studies have revealed that memory is selective and that our minds are very good at inventing images and memories to fill memory gaps. Our memories have been so incomplete that, in modern and advanced judicial procedures, the memories of witnesses and victims are not enough to declare the criminal guilty. Too many crass judicial errors were committed by relying on the feeble attributes of our minds. Of course, this has adaptive advantages. If reality must be reconstructed with what you have on hand, your mind uses feelings, expectations, and motivations to fill in the gaps. But, again, this is not acceptable as science. A classical example was provided by Allport and Postman more than half a century ago. They provided pictures of white and colored subjects to white college students. Although it was the white subject in the photo who showed a knife, about

half of the students reported that a black person was carrying a weapon. Clearly, relying on memory is not good enough. Notebooks are often used by scientists in lieu of memory. A scientifically-oriented student should learn how to keep a good notebook as soon as possible.

HOW IS SCIENCE DONE WITH A FAULTY MIND?

Scientific achievement is not possible without using your mind. Yet, we just showed that the human mind is rife with shortcomings and errors. How can science advance with such a faulty tool? Fine tuning of this tool is precisely the training offered by PhD studies. Each discipline has its own experiences, rules, tools, and theoretical background that allow science to progress. The specific method of training and using a faulty mind is the art of each discipline. Interdisciplinary forays often stumble when extrapolating ways of handling ideas and tools from one reality to another. The precise equilibrium between reliance on one's mind and skepticism is an art that has to be learned through practice, which varies slightly from discipline to discipline. There is no simple protocol for always doing good science everywhere. Thus, at least for now, there is no substitute for "hands-on experiments" in order to learn how science works. That is why good PhD programs offer extensive practical experience with the scientific method.

INTUITION, REDUCTIONISM, AND SYNTHETIC GENERALIZATIONS

Frequently, laymen confound the scientific method with reductionism. True, science draws heavily on reductionism or the simplification of complex problems. When trying to grasp complex phenomena, we must reduce and simplify the object of study in order to understand it. Yet, reductionism is not unique to science nor is it always appropriate. Reductionism is a limitation of science as it is necessary only due to the limitations of our mind. It is analogous to using a magnifying lens: it helps to see details better, but it impedes a more general view. On the other hand, the detail is lost when our mind tackles the general problem. This limitation can be applied to science and is important for understanding relevant features of our multidimensional world. Specialization in the field is required to dig deeper into the unknown, but leads us to even fuzzier general visions. Science will have to increase its investment in techniques to draw generalizations. These generalizations will help keep the myriad of detailed information as a coherent and consilient whole in order to cope with an ever larger body of knowledge. This is an important future challenge for science.

Until science is able to tackle these challenges successfully, human intuition will prevail when handling complex problems. Our intuition has been successful throughout human history, and we will changer it for another knowledge instrument only if the new one is much better. Intuition has its own limits and it is not a substitute for science. Better knowledge of the workings of human intuition will aid not only cognitive science but science and decision making in general. Until such an understanding is at hand, science must limit its analysis of complex problems and respect the use of intuition by decision makers working on complex problems such as politicians or managers.

It is important to keep in mind that many human attributes, such as strong will, motivation, beauty, endurance, and hard work, are fundamental to success in the modern competitive scientific environment, but they are not essential for the scientific method. In this same vein, abstracting, conceptualizing and drawing general conclusions are fundamental to model building. Model building is an innate action of the human mind. Many models, although not based on scientific proof and often demonstrably wrong, are sufficiently good for solving the everyday challenges of normal humans. That makes them useful, especially if they are very simple rules of thumb. But popular or ancient wisdom might be wrong and is no substitute for science.

Chapter Five

Science and Society

All of us who are concerned for peace and the triumph of reason and justice must be keenly aware how small an influence reason and honest good will exert upon events in the political fields.

—Albert Einstein

THE IMPACT OF SCIENCE ON HUMAN SOCIETY

In 2001, J Bradford DeLong (see Figure 5.2) gave an illustrative approximation of the impact of technological developments unleashed by science on the aggregate economic activity of humanity. Most authors agree that the human population on our planet has increased markedly as a consequence (or was the cause) of the "Agricultural Revolution." More food allowed more people to be fed.

The consequence of more people inhabiting the planet, each of them producing some kind of utility, increased the aggregate wealth of humanity, as was estimated by various authors using different methods and summarized by DeLong in the Figure 5.3.

In per capita values, humanity has not demonstrated a significant improvement after the agricultural revolution. In many occasions, world GDP might have been worse, as widespread famines and epidemics, unlikely to occur among hunter-gatherers, affected dense human settlements from time to time. It was only after the Industrial Revolution that humans were better off as individuals, as represented by various estimates in the Figure 5.4.

Figure 5.1.   Adapted from the color painting "Woman leaving the psy-choanalyst's office" by Remedios Varo.

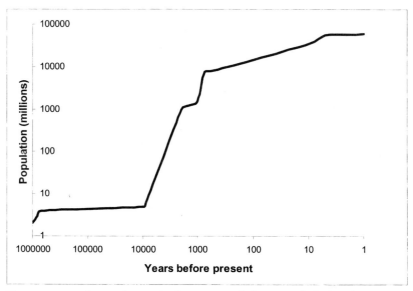

Figure 5.2.   Estimates for earth's human population growth during the last million years by J. Bradford DeLong, 2001.

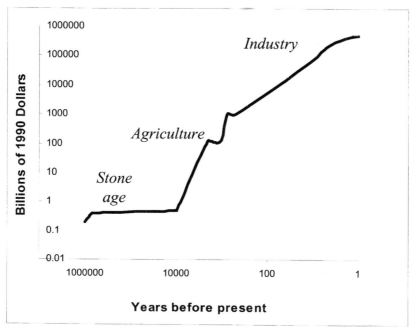

Figure 5.3.   Estimates of total material wealth enjoyed by humanity during the past million years by J. Bradford DeLong, 2001.

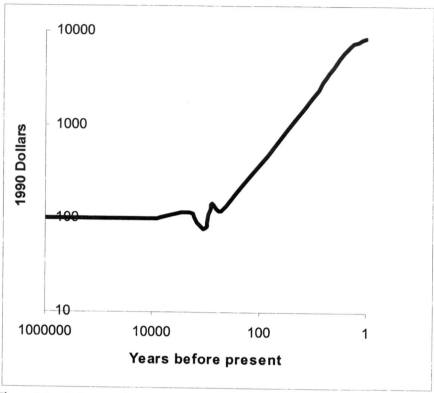

Figure 5.4.   **Estimates of total material wealth enjoyed by the average human being during the past million years by J. Bradford DeLong, 2001.**

   Two events are in view in a ten millennia time window: The emergence of the written word some ten thousand years ago and the emergence of experimental modern science a few centuries ago. Of these, science, which served as a trigger for the Industrial Revolution, was the most important event in the recent history of humanity. Certainly, it was the most remarkable event in the past millennia after the invention of the written word, as is schematically represented in the Figure 5.5.

   The Sumerian scriptures, written about five thousand years ago, reveal a *Homo sapiens* that was indistinguishable from modern man in terms of feelings, motivations, wit, religiosity, curiosity, and social values. Yet, modern experimental science did not exist then. It only emerged at the start of the Renaissance. I will call this the Galilean Revolution.

   The quest for knowledge in western history, after scriptures had been invented, as depicted in Figure 5.6, seems to be conspicuously heteroge-

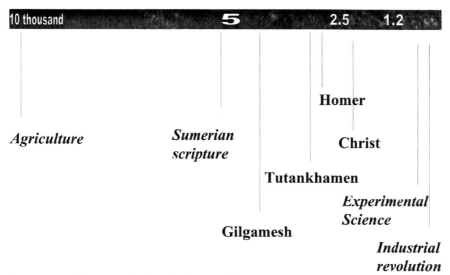

Figure 5.5. Diagram of a historic time windows.

neous. Rational thinking by authors, such as Pythagoras and Aristotle, seemed to have been displaced by pure mystical thinking over two millennia, until the arrival of the Renaissance. These thinkers did not separate religion from their models explaining nature. Religion actually displaced rational thinking for over 1500 years. The introduction of empirical verification as the final arbiter for alternative hypotheses was achieved

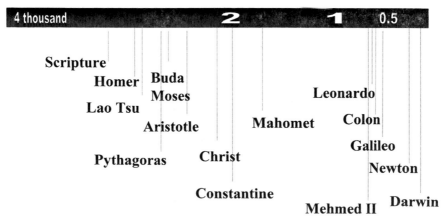

Figure 5.6. Diagram of a historic time windows.

only in the last couple of centuries, after the Christian Roman Empire collapsed.

Although Galileo Galilei was not the sole actor in this scientific revolution, he was the best communicator and the strongest advocate of this revolution. The revolution started when intelligent humans started to understand the world beyond our earth. Copernicus described the movement of the stars and planets in relation to the apparent movement of the sun. However, simple logic did not reveal the mysteries of nature. It was observations and experiments, performed with extraordinary beauty by researchers like Galileo that allowed us to gain new insights into our surrounding world. Once the scientific spirit was unleashed, it produced remarkable transformations, as depicted in Figure 5.7.

A positive feedback between science—the discovery of new knowledge—and technology—the design of new tools—began. The new knowledge allowed us to build telescopes and microscopes, machines and measuring devices, which in turn allowed us to discover new and finer aspects of reality. The new stream of unleashed technology profoundly transformed our culture and humanity in general. Thus, the emergence of modern science can certainly be viewed, together with the domestication of fire, the invention of projectiles, and the domestication of plants and animals, as a cornerstone of human evolution.

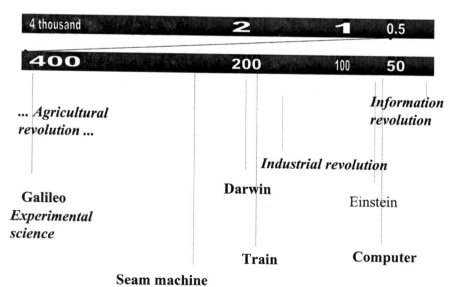

Figure 5.7.   Diagram of a historic time windows.

MODERN SCIENCE, SOCIETY, AND ECONOMIC GROWTH

Modern science, as we know it and as described here, is not an ancient concept. Galileo Galilei accepted the limits of human perception and the importance of observations and experimentation in establishing the truth of a theory. However, it was the formulation of Newtonian physics, later displaced by Einstein's physics that showed humanity the worth of provisional theories and the inscrutable workings of science in building hypotheses and theories, designing experiments, developing better predictions of nature, falsifying hypotheses, and building better hypotheses. The induction and deduction involved in the production of a hypothesis controlled by experiments and empirical observations is the soul of modern science.

Science does not solve all of the problems that we have as animals and as humans, nor is it easy to implement a rigorous scientific research program to solve many complex problems. Yet, science possesses qualities ascribed to democracy by the British Prime Minister Winston Churchill, when he said that, "Democracy is the worst form of government, with the exception of all others." He recognized that democracy as a system might have many defects, but it is certainly the least bad of all of the systems of government known to man. Science, and its daughter, technology, are endeavors with many defects and limitations, but they certainly represent the most efficient heuristic epistemological method known in the production of technological and economic progress. Science is the only known epistemology that has been shown to produce substantial benefits for humans and significant tangible improvements to the living conditions of the great majority of humans on this planet during the last centuries.

Societies that nurture science and that maintain environments where science prospers achieve economic prosperity (see books by Richard Florida, for example). This relationship between economic prosperity and science is evidenced in Figure 5.8, where the number of scientific publications per inhabitant of the various countries, as monitored by the Science Citation Index in 1992, is plotted against the Gross National Product in that country for the same year.

The relationship between science, technology, and economic production differs in various historical contexts. In pre-industrial societies, more science did not necessarily lead to more economic development. A meta-analysis of the relationship between science and economic development (see: Science, Religion, and Economic Development, published in the journal *Interciencia* in 2005) finds correlations between these two variables only among countries with an annual income per capita above 1,000 US dollars.

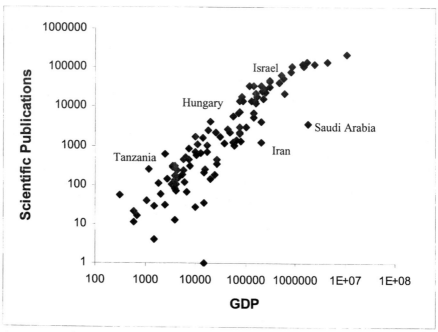

Figure 5.8.   Plot of the number of scientific publications produced by a given country vs. its total wealth produced (GDP) during 1992.

If we take a broad view of world affairs, we notice that humanity is in the middle of a profound and irreversible transformation. Some societies have experienced an Industrial Revolution and are prospering technically, economically, politically, and militarily.   They are referred to as the rich countries of the world. Most of them are in the Organization for Economic Cooperation and Development (OECD). In these countries, there are signs that a new revolution is starting: the "Information Revolution." In other societies, the Industrial Revolution is just starting or has not started. These societies still live in an "Agricultural Age" or in an age dominated by social and psychological attitudes adapted to hunting, fishing, and gathering food and other resources, which is more reminiscent of the "Stone Age."

An interesting fact about the heterogeneous spread of the Industrial Revolution among the various societies in the world is that it causes heterogeneous relationships between economic growth and human well-being. That is, science becomes important in industrial societies. At the same time, the transitions and socio-cultural changes from hunting and gathering to agriculture and from there to industrialization causes large inequalities in wealth among humans. Material inequalities, however, have a non-linear correlation with

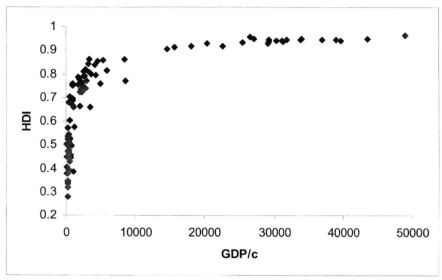

Figure 5.9. Plot of the Human Development Index for a country and its GDP per capita as reported by the United Nations Development Program in 2001.

well-being. This curious effect can be observed in Figure 5.9, where we plot the Human Development Index (HDI) for each county, as assessed by the United Nations Development Program (UNDP) in 2001, against its Gross Domestic Product (GDP) per capita.

This plot shows that an increase in the GDP of rich countries does not significantly affect the well-being of its citizens, as measured by the HDI, whereas, in poor countries, small increases in GDP are certain to significantly affect HDI. In transitional economies (indicated by the black arrow in Figure 5.9), which are suffering through the transition to industrialization, this relationship might vary, but they are intermediate between the situations described for the rich and poor countries. That is, in rich countries, increases in material wealth have a small effect on the well-being of individuals, whereas in poor countries, increases in wealth strongly increase human well-being.

The relationship between science and the creation of wealth is clearly evidenced only in the richer countries. Yet, this relationship can be seen even if we do not separate the different countries into categories. If, for example, we expand our analysis made at the beginning and compare the total number of publications produced by each country, as a measure of scientific productivity, with the wealth achieved by that country, we find that this correlation coefficient (0.88) is the largest among all possible features

Table 5.1.   Correlation coefficient between various indices of technical, artistic and scientific developments and economic development (GDP/capita) of a country

Variable of Countries	r-coefficient
Web hit's with Google	0.39
Total expenditure on education (% GDP)	0.49
Publications in Arts & Humanities	0.61
Publications in Social Science Index	0.64
Films produced	0.72
Number of years at which >90% of the population are enrolled	0.78
Publications in Science Index	0.88

tested that correlate with the wealth of a nation. In Table 5.1, we present a selected group of variables reported annually by the United Nations' statistics and their correlation with the wealth of a nation as expressed by its GDP (either in US$ terms or corrected by the Purchasing Power Parity for 2003).

Clearly, science is the cultural variable with the strongest relationship with economic development.

The important roles of science and technology and of services and information technologies in the achievement of industrial development are supported by numerous data. The precise mechanisms that scientific development employs to affect economic growth are still unknown, but there is no doubt that science and technology will be part of any modern economy. Many modern cultural forces are trying to dethrone science from the front of human progress. The consequences of such attempts can be very severe, as were the attempts of the Cultural Revolution in Mao's China, Cambodia's Khmer Rouge, or Afghanistan's Taliban to regress to an agricultural age in order to avoid the consequences of the Industrial Revolution. The elements presented here strongly suggest that the understanding of science, and its spread into most sectors of society, is not only an endeavor of academic interest, but is of utmost importance to construct ever more rational economic policies that favor the development of the majority in countries where huge proportions of their inhabitants still suffer from curable diseases and where famines are a fact of life. Understanding the phenomenon of science is fundamental to developing new disciplines that will allow us to understand humanity and its future.

Citing Fred Robinson, Chancellor of the Exchequers in London from 1823–1825, when addressing the Commons: "There is a principle in the constitution of social man which leads nations to open their arms to each other, and to establish new and closer connections, by ministering to mutual convenience—a principle which creates new wants, stimulates new desires, seeks for new enjoyments and, by the beneficence of Providence, contributes to the general happiness of mankind . . . and when we reflect upon the facilities which is given to its operation by the recent discoveries of modern science, and by the magical energies of the steam engine, who can doubt that its expansion is progressive, and its effect permanent?" That is, as expressed by Paul Johnson in the Birth of the Modern (1991), modern science and technology turn the luxuries of one generation into the necessities of the next. Clearly science has been the driver of innovation for the last few centuries and will continue to be so for some time.

SCIENCE AND ART

Artistic creativity has been a particularity of *Homo sapiens* and is thought to be fundamental to its society. Societies are often classified and recognized by the art that they foment and produce. There is a general consensus among most evolutionary ethologists that its evolutionary origin might well have been triggered long ago by sexual selection. Sexual selection is a force of nature that acts through the reproductive success of individuals. Males that succeeded in finding mates to reproduce were those with the best abilities to give a favorable impression to females looking for males with good genes that guarantee her and her offspring better survival. Due to strong competition between males, this favorable impression was achieved with ever more sophisticated actions and displays, testing the limits of the male's creativity and triggering an evolutionary driving force for evermore creative individuals. Important support for this hypothesis is that, in the animal kingdom, in many species, males perform songs and display complex behaviors to attract females to their love acts, whereas sound production by females is rare.

This creativity, although probably originating through sexual selection, has taken many forms and novel functions among modern humans. Two of these are art and science. Similar to two sisters, art and science have many similarities and some differences. Both are the product of human necessity for creating and innovating, but each has its particular method. The arts explore the world using intuition, reason, sentiments, inspiration, and work through trial and error; whereas science seeks to explore the world by incrementing the efficacy of search algorithms, creating hypotheses, and designing experiments to test them. It is not a surprise that we might identify elements of a hypothesis or experiment in the arts and extensive use of trial and error in science. It seems that there is only a difference in the degrees of relative importance. Yet, there are moments when degrees are important.

A fundamental difference between both sisters (science and arts) is that science recognizes the limitations of the human mind in perception, logic, and rational thought, thus seeking to explore the surrounding universe with rational extra-sensorial methods, such as experiments. The arts, in contrast, explore the limits of our sensations and dig deep into its properties, but keep human senses and perceptions as the primary focus for their works. That is, both try to balance the use of reason and intuition as intellectual working tools, but when contradictions between reason and intuition appear, the artist chooses the latter, whereas the scientist chooses the former. This difference turns out to be of the uttermost importance.

Both art and science are rather specific human characteristics. Traces of the incipient development of these behaviors which may lead to art and science have not yet been discovered among other animals. Its performance by robots and computers, however, should not be discarded. The keys to artistic and scientific creativity are a large dose of luck, which hints at the stochastic processes that support them. These will surely be unveiled in greater detail by future scientific research and might then be used by human made machines.

SCIENCE AND ETHICS

"I find it hard to accept that our deepest beliefs were set in stone by agricultural societies of the eastern Mediterranean more than two thousand years ago."

—Edward O. Wilson. (Consilience)

"Science is based on experiments, yet it reaches its results through conversations between scientists, discussing the meaning of experimental results."

—Werner Heisenberg (*Der Teil und das Ganze* 1969)

Ethics, or the epistemology of the good and the bad, is at the basis of every structured society. Despite its uneven history, some values seem to be permanent in human history, as they can be detected even in ancient times, such as those revealed through the Sumerian scripts. Others have suffered changes detectable even in times measured by a single generation, such as our valuation of abortion, homosexuality, and the production of human clones.

Ethics can be viewed as a young science that is still being developed. It deals with algorithms for human action that favor the individual and society. Sometimes, both aims are in contradiction, which raises fundamental moral questions. As with incipient disciplines, we must start by building a classification of ethical truth. For example, positive values detectable from Sumerian scripts that seem valid today include: Gentlemanship, Truth, Law, Order, Justice, Liberty, Rectitude, Sincerity, Piety, and Compassion; whereas the negative values include: the Bad and Evil, Lies, Anarchy, Disorder, Injustice, Oppression, Perversity, Cruelty, and Insensibility.

All classifications are arbitrary to some degree. The most simple of classifications, which serves efficient individual and social decision making, groups objects and phenomena into two categories: Good and Bad. Classifications using three categories are a little more sophisticated but continue to be sufficiently simple so as to serve as a base for dogmas, universal theories and religions. Simple classifications, if recognized as that, are useful for understanding intuitively complex phenomena.

If we accept that human knowledge is related to truth, and that both are an approximation of reality, we may recognize different degrees of knowledge and truth. Using thermodynamics as analogy, we know that as we reduce space and time, we reduce the total amount of entropy and increase the order or negentropy of the system. If we regard truth as a negentropic property, by reducing the scope in time and space in which we study a given subject, we will increase the degree of truth. Another way of increasing the degree of truth of a given statement or knowledge is by generalizing its scope. Laws, such as the universal law of gravitation, have no apparent limits in time or space for its scope of application. Yet, their implementation in concrete predictions is not trivial. This example shows that ethics may eventually become a discipline which can measure its objects of study quantitatively and which will eventually be able to develop predictable and falsifiable theories.

Ethics does not need to contradict science. It often is a seed for triggering the scientific study of social dynamics, i.e. the dynamics of individual algorithms as they relate to the social whole. This has been the case of the recent advances in our understanding of the roots of cooperation, altruism, shame and a sense of justice. It is highly probable that ethics will be a full fledged hard science in the future. This will only be possible with a much more sophisticated and rigorous body of human, behavioral, and social sciences. But quantitative and empirical science may be strongly limited in this task.

Science is not objective in the sense that it eliminates all interference from human emotions. On the contrary, science is dependent on the emotions of scientists. It is the human emotions, such as the drives for power, new knowledge, increased ego, or need for affection, that make science possible by humans. Science is objective only to the extent that it accepts the limitations of our minds and calls upon experiments and independent tests to decide between competing alternative theories. Science is an extra-personal, rather than an a-personal endeavor. As such, ethics can have strong influences on individuals performing science and thus on science. The ethical limitations that have been imposed upon science by conservative societies on bio-engineering, reproductive biology and social experimentation are examples of this negative influence.

THE SOCIAL SCIENCES AND THE EXCESSES OF POSITIVISM

Never allow yourself to take seriously the problems about Words and their meaning. That, what has to be taken seriously are the facts and our descriptions of them: theories and hypothesis; the problems they solve; and the questions they pose.

—Karl R. Popper

Applying insights gained from natural science to the social sciences is not new. The Chinese Confucius (551–479 BC), Shang Yang or Gonsun Yang (400–338 BC), Han Feizi (-233 BC), and Mozi (-429 BC) and the Europeans Hugo Grotius (1583–1645), Tomas Hobbes (1588-1679), Francois Quesnay (1694–1774), David Hume (1711–1776), Jean Le Round D'Alembert (1717–1783), Anne Robert Jacques Turgot (1727–1781), Joseph Louis Lagrange (1736–1813), Nicolas de Caritat or the Marquees of Condorcet (1743–1794), Adolphe Quetelet (1796–1874), and Auguste Comte (1798–1857) are just a few of a great many who attempted this endeavor. The extraordinary success of natural science in the XVIII and XIX centuries led intellectuals, which were highly concentrated in France at that time, to apply the techniques thought to be successful in the natural sciences to the social sciences. This historic phase of our intellectual development is often referred to as Positivism (see an excellent discussion by Friedrich A. Hayek (1889–1992), *The Counter-Revolution of Science Studies on the Abuse of Reason*).

As these simple extrapolations of analytical techniques from natural science to the social sciences where largely unsuccessful, post-positivist intellectuals discarded the whole exercise as useless, claiming that the study of humans must follow rules that are altogether different from the study of non-human reality. Today, however, we know that societies are complex systems and that extrapolations of analytical methods from simple systems to complex systems will invariably fail. Thus, the error of the Positivists was not the attempt to include the social sciences as another discipline of natural science by applying the same analytical tools to the study of both, but the failure to recognize that complex systems in general, be they human or animal, need more sophisticated tools for their analysis to handle the particularities of non-linear dynamics, irreversible processes, and emergent phenomena.

Although most of the earlier intellectuals that tried to bridge the human and natural sciences where aware of the limitations of their effort and did not claim irresponsible generalizations of their theories, later scholars, some of them with no serious knowledge of science, such as Saint Simon, made exaggerated and unfounded claims that led to serious problems and a rejection of any attempt to bridge the gap between both sciences. Some of the problems can be summarized as follows.

- Lack of humility in recognizing the limits of science.
- Lack of objective subjects of study and mixing of subjective assessments with objective measurements. Subjects and the subjective will affect the objects. For example, expectations, traditions, and values affect prices which affect objective measures of an economy.
- When classifying and referring to objects, we normally assume that the other possesses a mind that is similar to ours and classifies his surroundings similarly. This, however, is an assumption that does not hold with complex social phenomena. People might interpret the same act or situation with different values, such as classifying the Latin American guerilla Ernesto "Che" Guevara or Carlos "The Jackal" as heroes or communist cranks.
- Reductionist approaches to the study of social phenomena, without adequate attention to different time windows and spatial perspectives, leads to bitter disputes between biologists, sociologist, economists, and physicists. We are all exploring the same elephant in the dark and touching different aspects of the same beast.
- As Adam Smith remarked, we humans constantly promote goals that we originally never thought to favor. That is, the workings of unpredictable consequences in complex systems. The mix between want and object obscures any attempt at a scientific study.

These shortcomings and flaws of Positivism are not final, as they might be overcome with the advent of more sophisticated social sciences. Yet, there is a recurrent question: Are the social sciences an empirical science?

Many human efforts were directed at demonstrating that the social sciences and natural science are fundamentally different due to the fact that their objects of study are different. It is argued that the social sciences deal with human subjectivity and thus, can not use the methods of natural science with the same rigor. Most of these works, including that of Friedrich Hayek, used definitions and conceptualizations of science that were based on the form of natural science, i.e. mathematical and other abstract reasoning, more than on the principles of science, i.e. acknowledgement of the limitations of the human mind and the need for falsifiability or independent empirical support for theories.

The development of disciplines to study complex systems, the enormous increase in calculating power by computers, the ever increasing amount of available data on larger amounts of human activity, leads us to think that the rise of a more empirically-based, falsifiable study of human and social issues is on the horizon. As the aims of social science include many complex phenomena, it is reasonable to think that it will take longer to develop and mature as a science than others that study the interactions of simple particles, such as physics. We should view the social sciences as incipient or young disciplines with a very interesting future. They suffer from many non-scientific historic influences, but they have the potential to develop into a full-fledged hard science. Recognizing the fundamental qualities of science in general will help to guide our efforts in further development of these important academic disciplines.

Chapter Six

The Dangers Ahead

The further development of science and its positive impact on human society is not graciously or automatically guaranteed. Many things can go wrong. An exploration of possible dangers and pitfalls will surely not damage any positive future prospect for science and humanity.

NEG-ENTHALPING

One biological strategy to improve the odds of transcendence consists of filling the available spaces or niches in the ecosystem to improve the chances of a longer permanence in time on this planet. Systems which achieve this reduce the order of the system by decreasing the available competitors. In thermodynamic speak, they increase the entropy of the system, decrease the available space for competitors, increase their own resilience, and become more conservative. Thus, the aggregate force used (work) or the degree to which niches are filled and the finesse of its occupation are indices of the age or degree of evolutionary maturity of a given system. This expansionary cumulative force or work is homologous to what we know in thermodynamics as enthalpy. The opposite force is termed *negenthalping*, which restrains the expansionary drive.

When applied to the possible dominance of science in society, an increase in entropy means that science will try to fill all available niches in human thought and culture. That might not be a good idea and some form of neg-enthalping would be advised. Human biology must release many irrational drives and instincts, guided by art, religion, sports, and other means, to allow for harmonic function of the organism. Science should not interfere with

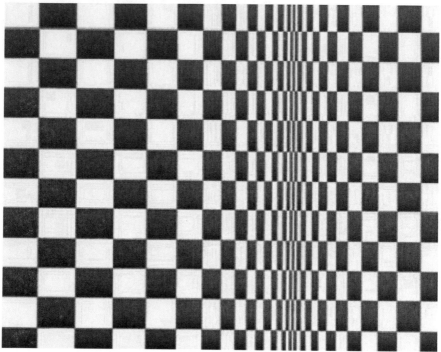

Figure 6.1.   Optical illusions.

those attributes, it must be restrained from invading their territory unless it wants to awake the ire of humanists and the common people and risk eventual extinction.

Thus, over-rationalization might be misplaced in many circumstances. The law of the unintended consequences applies to science. As Friedrich Hayek put it, civilization has developed and is based on many institutions that were not built on rational principles. Rationalists that do not acknowledge the limitations of our minds and methods of analysis and who ignore the inborn and cultural necessities of humans are bound to produce damage. The more power they get, the more damage they might cause.

DOGMA, MYTHS, AND RELIGIONS

When the human mind discovers new objects, phenomena, or relationships the brain secretes special neuromodulators that accelerate learning of whatever we happened to be learning or experiencing at that moment. This positive-reinforcement is driven exclusively by our emotions and is not amenable to rational manipulations. Therefore, it is the source of many myths.

When we are unaware of its workings and if we do not know the limitations of our mind and thoughts, we end up building abstractions and modes of our surrounding reality that can be systematically disproven. Our emotions will still cling irrationally to these models to form the basis of dogma.

These dogmas differ from scientific paradigms in that they can never be falsified and thus, will keep influencing human society for long historic periods of time. When dogma and myth extend into a large part of society and lead to social institutions based on them, religions emerge.

One important consequence of dogma and religion in human societies and individual minds is that it paralyzes social creativity and individual innovative thoughts. Religion has not been associated with creative explosions of knowledge in human history during the last five centuries. New insights into

Figure 6.2. Giordano Bruno (1548-1600).

our surrounding world have not always been appreciated and were actively op-
posed by religion, as vividly exemplified by the burning of Giordano Bruno
(1548–1600) because of its cosmological views, by the Catholic Church.

Even the fundamental insights of Nicolaus Copernicus (1473–1543) and
Johannes Kepler (1571–1630), which were brilliantly exposed by Galileo

Figure 6.3.   Galileo Galilei (1564–1642).

Galilei (1564–1642), were hindered and censured by the church until the end of the last century. This is the most conspicuous case of religion arresting the advancement of science, but many other examples exist.

New censures of innovative science by religious dogma can be found today in many places and instances. Research on reproductive biology suffers from multiple legal, moral, and violent actions to stop human cloning. The study of evolutionary biology suffers from the reticence of fanatic religious groups to accept biological evolution as a reality. Research into social dynamics and the spread of sexually transmitted diseases is hindered by several religions that fail to accept homosexual and other sexual behaviors as practices that merit our understanding. The areas of conflict between dogma and religion with science seem to be particularly intense today in areas regarding human reproduction.

Dogma and religion had a strong adaptive advantage in early human society, as they helped to form strong social bonds and provided a fabulous social force for the defense of the society by creating strong social cohesion. Many primitive hunter-gatherer societies survive today due to these extraordinary forces. These same forces, acting on modern global societies, may cause terror, violence and wars.

Myth, dogma, and religion are not the only forces opposing science. Other features of modern human society seem to collide with science as well.

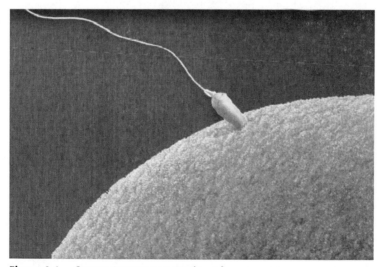

Figure 6.4.   Spermatozoon penetrating a human ovum.

OTHER ENEMIES OF SCIENTIFIC PROGRESS

A series of rational academic approaches to reality are often confused with the scientific method. These forms of rational thinking are clearly un-scientific, but they have some properties that make them appear to be sophisticated academic thought and are thus often accepted as serious. Some of these forms of rational thinking have been classical enemies of science, even before the Renaissance, and are still used in some contemporary academic circles. The most important of these, besides dogma and religion, are the prevalence of authority over facts, the strength of ideas that force data to adjust to them, and a tendency for anthropomorphosis that force human form on all objects of the world. The basis underlying these dogmas is that reality has already been revealed by Holy Scriptures, heavenly words, or otherwise, and that our rationality might serve only to explain real life using the accepted scriptures, knowledge, or dogmas.

Academics often base their digressions and analysis on the opinions of other academics. Dialogue, discussion, and criticism with other opinions have advanced true academic knowledge. It is the theory of Dr. X in opposition to the thought of Dr. Y that enriches academic discussions. Often, the more senior authority or the more convincing or emphatic argument will win the discussion. That is, ideas are often thought to be more important than reality. It is a strong view among humans that a beautiful idea can not be wrong and, therefore data and events from the real world have to be adapted, coerced, and molded to fit the idea. This attitude does not need empirical facts to advance knowledge. The consequence of this attitude is that many philosophical discussions that were held centuries ago do not differ much from those of today.

Anthropomorphism or the projection of the workings and structure of the human mind on the surrounding reality is an appealing form to rationalize our experiences. It comes naturally as it is an extrapolation of our selves to reality. Seldom, however, is reality compatible with the structure of our mind and anthropomorphisms have led to important false theories in the past.

Authoritarianisms and conservatism will stop science. For example, the Confucian tradition of respecting customs and hierarchy has cast a long shadow over the scientific activity of China, Korea, and Japan until very recently. Authoritarian rule and political conformity in the past decades hampered the creation of an environment that fosters individual creativity. Deference to authority and to existing paradigms is a major barrier to scientific breakthroughs. Although science education in China and Japan is extensive and rigorous, their scientific productivity, as measured by scientific papers, is lower than expected for the amount of human and financial resources poured into science. This suggests that it takes more than money and classical education to

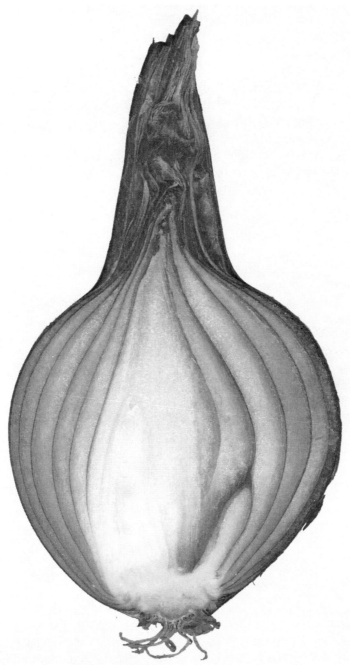

Figure 6.5.   Onion: *Allium cepa.*

cultivate scientists; students should be inspired to pursue knowledge for their own sake, independent from any authority, to acquire the habit of raising questions. Skepticism and iconoclast attitudes need to be fostered.

Technology is the main driver of modern science. It allows us to improve our senses and extended the functions of our organs, to expand our mind, and to perform ever more sophisticated experiments. Technology also creates filters between us and reality. It colors the things we see and molds our empirical experiences. Thus, technology, or rather the excess of it, might hinder deeper insights into reality and hinder science. Awareness of the working and limits of our technology will help us overcome this limitation.

Another important enemy of future scientific progress might include social values that are easily ingrained in modern culture. These values can have the following consequences:

• A lack of motivation for studying scientific disciplines, as is increasingly the case in advanced industrial societies.
• A lack of imagination in those doing scientific research, as scientific disciplines become even more specialized and competitive and are driven by economic motivations.
• A lack of scientific education in large parts of society, as the internet and television causes young students to have smaller attention spans and to limit their thought to processing simpler bits of information.
• A search to solve non-existent problems or analyze imagined facts. The focus turns to non-scientific questions driven by politicians, journalists, and other persons in power with a very rudimentary knowledge of science.
• Perference for charlatanry, barber-shop politics, and the search for simple solutions to complex problems, as they address inborn behaviors, instincts, motivations, and sentiments that are unconsciously driving most humans.

The example of the onion might illustrate this point. An object with so many protective layers must hide something important inside it. Go look for it!

MERITOCRACY VS. DEMOCRACY

The Limits of Democracy

Politics does not pardon science and science policy is important to scientific progress. Thus, management systems that successfully managed a republic are being applied to the management of science. Through this logic, democracy is being applied with varying degrees of success to scientific organizations, funding agencies, and academic associations. Democracy is not a panacea; it has limits.

In the quest to escape the shortcomings of authoritarianism and autocracy, a variety of alternatives have been employed. Some of these alternatives were completely unsuccessful, while others were very successful. Democracy is by far the dominant organizational structure of society in the XXI century. Democracy has many advantages *vis a vis* other organizational structures, especially in the coordination of a nation or when used as a basis for a Republic. It is considered by all international organizations and declarations of principle to be the most advanced system of government.

Yet democracy, when applied to scientific organizations, universities, research institutes, research laboratories, or even scientific activities, may become a strong limitation for scientific progress. No law of nature has been discovered or described by the popular vote. Democracy has not been a strong force for the advancement of science. Meritocracy and the force of empirical evidence are the building blocks of successful scientific organizations. Extrapolating the critique of democracy by Irving L. Janis in *Groupthink,* to the scientific environment, we might pinpoint the following dangers of a wide application of democratic methods and attitudes to science:

- The illusion of invulnerability
- The illusion of unanimity
- Suppression of personal doubts
- Self-appointed mindguards
- Docility fostered by suave leadership
- The taboo against antagonizing valuable members

These and many other limitations to democracy make it an inappropriate tool to help decide the truthfulness of the laws of gravity or the convenience of accepting or discarding evolutionary theory. What are the alternatives?

Reflections on Meritocracy

Meritocracy is a government ruled by those individuals who have shown merits or exceptional abilities in solving problems and performing activities relevant to the society to be governed. It should not to be confounded with Aristocracy or Autocracy, or with a lack of Democracy. Meritocracy values success and favors those who are more able to achieve it. Because merit depends on the effort invested by the individual and their degree of motivation and willingness to learn, it is not limited to an individual's genetic or cultural capacity. The amount of effort required to achieve a given level of "merit" varies among individuals which makes meritocracy a powerful form of organization. Meritocracy has the following goals:

• To value merits, favoring the more valuable over the less valuable
• To value effort more than opportunity
• To value the good in others more than their mediocrity
• To recognize that we are part of progress and that we are not always at the front of the wave
• To value productivity more than titles, honors, ranks, and positions
• To recognize the weaknesses in ourselves and the strength in others as an unavoidable step in the process of our growth.

To favor the average, the mean, the common over the valuable, the exceptional, the shining, is "Mediocracy" and is favored by the mediocre. The mediocre minimizes effort, whereas the meritocrat maximizes the quality of his actions.

A direct relationship between science, development, and meritocracy seems to exist (or between underdevelopment and a lack of meritocracy) at the level of a family, company, country, and culture. Underdevelopment and socio-economic backwardness are often consequences of a lack of meritocracy, which in turn, is a consequence of an excessive prevalence of personal ambitions of "Big men" over the common good. Societies that lack meritocratic systems harbor individuals that blame outside forces for the lack of socio-economic progress, rather than looking to themselves as a potential source of progressive force. Backwardness seems to be strongly correlated with a lack of meritocracy and a low valuation of science. A view from the opposite side of the spectrum shows that higher levels of socio-economic development with a higher prevalence of science correlated with a wider use of meritocratic systems. Examples can be found by comparing highly competitive research universities in the USA (and some emerging in Asia) with state nannied universities wrapped in read tape in the third world and in many parts of modern Europe. Only meritocratic institutions are at the top of the world rankings.

• These relationships produce the following intuitive paradoxes:
• The wiser you are, the humbler and more tolerant you become
• The more you learn, the more you recognize your ignorance
• The less you have studied, the more you think you know
• The more ignorant you are, the more fanatical you might become
• The more fanatical you are, the more likely it is that you will reach a position of command in your society

- The more democratic a non-meritocratic society is, the more arrogant and ignorant the individuals will be that dominate the society, leading to a mediocratizing democracy
- The emptier the carriage, the more noise it makes when rumbling over the patchy road

Physical laws and socio-dynamic rules indicate that if we favor the average, we will not progress. By selecting the average characteristic, we will not achieve a shift in the frequency of that characteristic in the population. As recognized by evolutionary population biology, mediocracy stops evolution. Progress implies a shift in the average and a preference for above average solutions. Thus, democracy will achieve social progress in a society only if it allows or foments meritocratic components or features in its system that might help society overcome its weaknesses.

This certainly is true of successful scientific institutions where democracy has been displaced by a meritocracy. Peer review in high impact scientific journals and grant distributing agencies, appointment of senior positions in research institutions based on merit, scientific careers of individual researchers strictly driven by scientific productivity, etc., are all meritocratic inroads into modern society. Yet, too many modern scientific institutions are strongly handicapped by either excessive authoritarianism or unchecked democracy. Apparently, implementing a functional and smoothly working meritocracy requires time and tradition.

Empirical proof is required to help us identify the most efficient method of organizing humans for the quest of science. Scientific productivity and inventiveness, or lack thereof, will help determine which of the various organizational structures that are used for scientific institutions today should prevail in the future.

Social Progress: Fiction or Reality?

Will the current trends that we have experienced in the last few centuries continue? Will science and social progress continue at an accelerated pace? Drawing from mental models of thermodynamics and extrapolating from studies of complex multi-component processes in the physical science may teach us that:

- In non-linear open systems, equilibrium dynamics is not possible. No static solutions are stable. Only stationary states are achieved in dissipative, far-from-equilibrium systems. Thus, no simple stable solutions will apply to the dynamics of complex social processes. We should expect change to be the only predictable phenomena.

- Information gathering for appropriate decision making has costs. Thus, limits to rational decision making will always exist. These limits will be more noticeable in the future as costs for more information will only increase.
- Maintaining equality in societies has costs. Thus, human societies will always profit from human inequalities, which will continue to be the driving force of future change.
- Maintaining creativity is in conflict with equality. Thus, equal societies, aiming for an increasingly just society, will sacrifice creativity. We can not discard the coming of a new "dark age," where creativity will not be valued as highly as it is today.
- Happiness is a meta-stable, out-of-equilibrium state maintained by the balance between two forces: achievements and aspirations. Thus, society will always change its form and structure. In the words of complex system theorists, the future promises many walks at the edge of chaos.

If we apply these insights to our view of future scientific progress, we find that scientific progress is possible only with social imbalances. Humans need to feel some kind of unsatisfactory experience or sense of frustration to awaken the inborn drives that motivate scientific research. Wars have been a strong driving force for scientific innovation in the past. Our fight against illness and death drive contemporary research in biomedicine. What is required to keep the engine of scientific progress running?

Many such questions are in need of clarification. How does the variance in inequality correlate with evolutionary processes? How does research efficiency affect the concentration of resources *vs.* dispersion of accumulated resources? What favors creativity and what helps to develop skepticism? What cultural traits hinder progress and are they related to the development of scientific skills? What is the relation between science, culture and religion?

PATENTS, CREATIVITY, AND A SCIENTIFIC ATTITUDE

In his autobiography, Karl Popper affirms that creative thought is a combination of intense interest in a subject or problem together with a highly developed critical attitude. A lack of one of these attributes will stop the advancement of science. Future societies might be less able to motivate people to engage in scientific enquiry, because political correctness and democratic demagoguery may limit the critical attitude of its citizens, damaging science in the process.

Ideas are cheap—very few people pay for ideas—because ideas are abundant. Yet ideas are fundamental for change and progress. The challenge for success is selecting the right idea and filtering out the bad ones. Science shows us how to do this. But you need motivation to do this. Traditionally, motivation for scientific research has been individual curiosity. It will always be important. Increasingly, a strong source of motivation for scientific research is economic: the patent. With patents, ideas are assigned a monetary value and their implementation is no longer free.

The discussion of the suitability of patents to stimulate new scientific research, and their role in hindering new scientific and technological developments is an old one. No amount of verbal rationalizations and legal discussions will show us the solution to this dilemma. Scientific analyses of practical implementations of the patent system should guide us in our determination of the equilibrium for maintaining a strong driving force for scientific development. There is not necessarily a uniform solution for all fields of science. Why is the duration of legal protection of an idea that takes a few hours to develop and implement, as might be the case for many software patents, the same as that of a chemical, pharmaceutical, or mechanical process that takes billions of dollars, several years, and hundred of people to develop? The answer certainly has more to do with politics and special interest groups than with the optimization of scientific progress. The wellbeing of future societies will depend on the adoption of a patent system that stimulates science and innovation. To find an efficient design for such a well functioning system is a challenge of our contemporary world.

Chapter Seven

Natural Selection and the Scientific Method

May be we can gain some insight from the life of an important philosopher of science. I call this story: *When ideas are above reality.* Around 1950 in Ithaca, the philosopher Stuart Brown wrote about one of the most influential philosophers of science: Ludwig Wittgenstein: "Ordinarily, he would refuse the offer of a ride. But one afternoon, after it had begun to rain, I stopped to offer him a ride back to the Malcolms. He accepted gratefully, and once in the car asked me to identify for him the seed pods of a plant which he picked. 'Milkweed' I told him and pointed out the white sap for which the weed is named. He then asked me to describe the flowers of the plant. I failed so miserably that I at length stopped the car by a grown up field, walked out and picked him more plants, some with flowers and some with seeds. He looked in awe from flower to seed pods and from seed pods to flowers. Suddenly he crumpled them up, threw them down on the floor of the car, and trampled them. 'Impossible!' he said." (From: *Ludwig Wittgenstein*, by Ray Monk. Vintage, London 1990, page 553).

Not all scientist and philosophers are conscious about the central role of empirical evidence in pursuing the scientific method. My insistence in focusing on the centrality of the experiment and empirical validation in understanding Science rises, as a natural consequence of applying theoretical evolutionary biology to human society. Nature has devised a heuristic or method to explore viable solutions in a world of infinite possibilities and unknown constraints: evolution by natural selection, which works by blending a system to manage information, with creativity and serendipity, and empirical validation; the same blend as that defining the scientific method. With Science, human civilization has started to understand and apply successfully the princi-

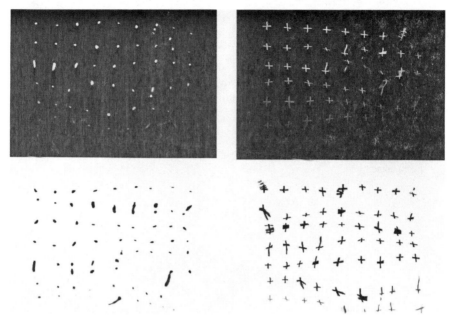

Figure 7.1. Plastic art by Magdalena Fernandez.

ples of biological evolution, even to Science itself. No other more powerful heuristics is available to humanity today.

Let's suppose that a successful heuristic system (or society) is ruled by a series of norms (laws): a finite group of actions that should be avoided and another finite group of actions that should be promoted. Regardless of the size of each of the groups, it is always possible to imagine an action in a given moment that is not regulated by the existent norms. Yet we can imagine a mechanism by which new norms are created each time a new action, not regulated by previous norms, becomes necessary or possible. The design of such mechanisms would seem an ideal aim of science or human ethics. In practice, our experience tells us that these mechanisms behave fundamentally in one of three ways (based on Lezak Kolakowski):

1. They are consistently consistent, leading eventually to fanatism and authoritarianism
2. They are consistently inconsistent, resulting in anarchy and chaos
3. They are inconsistently inconsistent, or, what seems to be equivalent, inconsistently consistent, reflecting what is apparently happening with the history of ethics.

Although Lezak Kolakowski formulated these alternatives semantically in a somewhat different form—he proposed 4 alternatives, separating alternative 3 in two -, I would like to call them the three Kolakowsi solutions to ethic norms. It is fascinating to note that Kolakowsi's third solution is also the one preferred by biological evolution, allowing random mutations to change old adaptations but at the same time, maintaining a strict order and tradition, both in an exquisitely fine equilibrium, thanks to natural selection or empirical validation.